The
Amana®
Radarange®
MICROWAVE OVEN
Cook Book

All recipes contained in this book were created, tested, or approved by the staff of the Amana Refrigeration, Inc., Ann MacGregor Home Economics Department.

PRECAUTIONS TO AVOID POSSIBLE EXPOSURE TO EXCESSIVE MICROWAVE ENERGY

(a) Do not attempt to operate this oven with the door open since open-door operation can result in harmful exposure to microwave energy. It is important not to defeat or tamper with the safety interlocks.

(b) Do not place any object between the oven front face and the door or allow soil or cleaner residue to accumulate on sealing surfaces.

(c) Do not operate the oven if it is damaged. It is particularly important that the oven door close properly and that there is no damage to the: (1) Door (bent), (2) hinges and latches (broken or loosened), (3) door seals and sealing surfaces.

(d) The oven should not be adjusted or repaired by anyone except properly qualified service personnel.

Dear Friend,

We are pleased to share with you the fun and ease of microwave cooking. Although it may be a new experience, microwave cooking is easily learned.

Whether you are single, a family of two, a new mother, or a large family, you'll find your Radarange Microwave Oven to be a real asset. Cooking time is greatly reduced, and electrical energy usage is decreased.

Through our travels and consumer programs, we've found the need for a variety of recipes to fit today's numerous lifestyles. The Amana Radarange Microwave Cook Book offers a collection for those who are true epicures, as well as for those who prefer "standard" recipes. "Quick-to-do" recipes are included for the working woman, busy homemakers, and bachelors. If you are cutting calories, check the special Low Calorie section. All ages can enjoy "Cooking for One or Two".

We're certain that your Radarange Oven will become "your best friend", by saving personal energy and allowing time for leisure. As you read through these recipes, you'll find "Micro-Tips" and other useful hints to help you understand microwave cooking.

We at the Ann MacGregor test kitchen hope you will share our enthusiasm for microwave cooking and let the Radarange Oven help you to prepare meals in minutes.

Ann MacGregor

Contents

1
Introduction

MICROWAVE OVEN COOKING TERMS

ARCING: A static discharge of electricity causing a spark. This usually occurs between separated particles of metal, such as a metal twister for plastic bags, gold trim on a dish, or a metal utensil almost touching the side wall of cooking compartment causing a spark.

ARRANGING: Suggested placement of several items of the same food in the Radarange Oven to produce the most satisfactory cooking results for the particular food. Potatoes serve as one example.

BASTING AND/OR DRAINING COOKING JUICES. There are two methods we recommend: (1) Drain accumulated juices which, if left in the oven, would slow the cooking process. Reserve and use a small amount to trickle over the surface of the food, if desired. (2) Trickling a small amount of liquid that was reserved at the time the recipe ingredients were combined over the food surface.

BROWNING: The change in outside food color which occurs in cooking. The degree of darkening relates to (1) the length of cooking time, (2) the type of food and (3) the surface temperature. Noticeable color change begins in roasts and poultry as the food's surface temperature increases. For small, short-term cooking items, browning dishes are used to obtain a color change.

BROWNING SKILLETS: Special microwave energy-absorbing dishes, which after preheating, produce heat for browning food by surface contact.

BURSTING: The build-up of steam or pressure in the food product that causes the exterior surface to split open. This may occur in foods such as apples, eggs, poultry and vegetables which are enclosed in a skin, shell or membrane. In order to prevent bursting, it is necessary to puncture the skin or membrane, or remove the shell.

COOK: To cause a temperature rise and physical change in food.

COOKING TIME: Time required to heat or cook foods to a serving temperature in the Radarange Microwave Oven.

COOKING TECHNIQUES: Methods to produce good cooking results. See: Arranging, Basting, Cooking Time, Covering, Puncturing, Rearranging, Rotating, Standing Time, Steaming, Stirring, Turning and Undercooking.

COVERING OR WRAPPING: Placing a glass, plastic cover or bag securely over a dish, thus retaining steam. Steam retention allows for a more even distribution of heat, prevents dehydration of food, and helps prevent spatters on the walls of the cooking compartment. Plastic wraps should be pierced to allow some steam to escape. Do not cover casseroles or dishes unless stated in the recipe.

DEFROSTING (THAWING): Method of applying microwave energy to frozen foods with short intervals of power, during which time heat is allowed to distribute itself throughout the entire food. Food is defrosted with a minimum of cooking.

OVERCOOKING: Cooking too long causing drying out, toughening of foods, separating of sauces, and possible hard spots in some foods. Careful timing on short cooking items is necessary. Overcooking is not always visible.

PUNCTURING: Breaking the skin or membrane of foods such as vegetables or eggs, which allows the steam to escape and prevents bursting. This also applies to puncturing of pouch cooking bags.

2
Introduction

REARRANGING: Changing the position of several items, such as potatoes or chicken parts, for a more even cooking.

REHEATING: Bringing cooked foods to serving temperature.

REHYDRATING: Replacing water or other liquid missing from freeze-dried foods.

RESTING: See **STANDING TIME**

ROTATING: Turning dish or cooking container one-fourth to one-half turn for more even cooking. Unless otherwise stated, all recipes in the Radarange Cook Book refer to a half or 180° turn.

STANDING TIME (RESTING): The amount of time suggested, either during or after cooking, which will allow the heat in the foods to equalize or spread to the center of the foods. Sometimes referred to as carry-over cooking.

STEAMING: Cooking in covered utensil. Lids should be left on during standing time to assure best results.

STIRRING: Manual movement of foods within the cooking dish with a spoon or suitable utensil to distribute the heat generated by the microwaves. Casseroles, hot dishes and sauces are often stirred during the cooking process.

TURNING: Inverting foods such as roasts during the cooking cycle. Unless otherwise stated, the Radarange Oven Cook Book refers to a half or 180° turn.

TRIVET: An inverted saucer or casserole cover used to elevate foods such as poultry and roasts above the cooking juices.

UNDERCOOKING: Cooking to a less-than-done degree, visually or physically, to allow for standing time to complete the cooking.

UTENSIL: Any dish or container that is used for Radarange Oven cooking.

Utensils recommended for use in Radarange Oven:
The following materials are acceptable for use in the Radarange Microwave Oven because they allow energy to pass through and penetrate into the food:

1. Heat tempered, oven-proof clear glass utensils without metal trim.
2. China, ceramics, some glass ceramics, opaque glass or pottery, without metal trim or metal base glaze.
3. Plastics such as plastic wrap or freezer containers should be used with care and usually only for warming food. Be certain plastic containers are labeled "may be placed in boiling water" or "dishwasher-proof". Not recommended for greasy foods such as bacon. Hot grease melts plastic.
4. Paper, such as plates, cups, napkins, towels or waxed paper, except for large food items having a high fat content.
5. Wood or straw, for very short periods of time.

Utensils NOT recommended for use in Radarange Microwave Oven:
The following materials are not acceptable for use in the Radarange Microwave Oven because they reflect, absorb energy, distort or melt from the heat of the food. Follow instructions because the incorrect use of some metal trimmed utensils may cause arcing or harm the magnetron, oven cavity or dishes. Also, inferior cooking can result.

1. Metal utensils, meat thermometers or utensils with metal parts.
2. Dishes with metal trim, band or decoration.
3. Dishes not recommended for use in the Radarange Microwave Oven include Centura® Dinnerware, Corelle® closed-handled cups by Corning™ Glass Works, or Melamine® plastic. Composition of material and glaze in these items makes them unsuitable for

cooking with microwave energy.
4. Aluminum foil, except as recommended in our recipes.
5. Wooden cooking utensils which may dry during long cooking periods.
6. Waxed or soft plastic containers because they can be melted or distorted by hot foods.

UTENSIL TEST: A simple exercise to help determine whether to use a particular dish or container in the Radarange Microwave Oven:
1. Place a glass measuring cup of water and the dish to be tested in the Radarange Microwave Oven.
2. Set the timer for 1-1/4 minutes. Test results: If the water is very warm and the dish is cool, you may use the dish. If the dish is slightly warm around the edges, use it for short term cooking only. If the water is cool and the dish is hot, DO NOT USE THE DISH.

BROWNING SKILLETS

The Amana Browning Skillets are footed microwave dishes with covers made by Corning™.

The skillets come in two sizes: 9-1/2 inch and 6-1/2 inch.

Both have a tin oxide coating on the bottom exterior. The tin oxide has the ability to absorb the microwaves and turn them into heat. When used as directed, the temperature will not rise beyond 600° . The covers are Pyrex brand heat-resistant clear glass.

When preheated in the Radarange Oven, the skillets can be used for browning meat, fowl, fish, sandwiches, pancakes and also for frying eggs.

The skillets also make excellent dishes for casseroles, vegetables and other foods. It is "freezer-safe" and can go directly from the freezer to the Radarange Oven without first being defrosted.

You may order them and other Radarange Oven accessories from your Amana dealer.

HEATING OF FOODS

Volume

As the volume of the food is increased, the time required to cook or heat the item increases proportionately. If twice the amount of food is placed in the range, it will take almost twice as long to cook. For example, if 1 strip of bacon requires 1 minute, two strips of that same size will require 1 minute, 45 seconds.

Starting Temperature

This is one of the most important influences on the required heating time of food prepared in the Radarange Microwave Oven. Each temperature degree that the food item is to raise must be supplied with a definite amount of energy. Lower initial starting temperatures mean a larger energy requirement and a consequent increase in cooking time. Therefore, refrigerated foods require longer cooking periods than foods stored at room temperature.

Density

Light, porous items, such as breads and pastries, will absorb microwave energy faster than a more compact moist item of the same weight, thus requiring only seconds to thaw or heat.

Moisture Content

Water, and water-laden foods, require more energy to reach a given degree of temperature than like items with a lower moisture content. The higher the moisture content, the longer the cooking time.

Fat and Sugar Content

Foods containing high fat and sugar levels heat very quickly. Those foods with lower fat and sugar levels tend to require longer cooking times.

DEFROSTING

You can defrost foods very quickly in the Radarange Oven. Defrosting takes place by alternating heating and standing times. This allows heat to be conducted to the center of the food, resulting in an even defrost with no outer-edge cooking. You may use the Automatic Defrost feature to accomplish this.

It takes about 5 minutes to defrost a pound of ground beef, or about 80 minutes to defrost a 16-pound turkey. The product will retain many of the juices that would normally be lost through the conventional "let it stand for hours" method.

Thawing may be followed by heating to serving temperature, or by primary cooking.

Hints for Defrosting

- Porous items like breads, cakes or pies defrost quite quickly.
- Dense items like meats and casseroles defrost quickly on the outside, but require a longer time for the center to defrost. With these foods, standing time is necessary to allow the center to defrost without cooking the outside portion.
- If cooking begins, decrease the length of the microwave periods and increase the standing period. Reposition the item in the Radarange Oven.
- Keep food wrapped in plastic or paper until ready to separate or rearrange the food.
- With meat products over 5 pounds, allow an additional standing time after defrosting to assure more even cooking results.

COVERINGS OR WRAPS

Most foods heated in a Radarange Oven should be covered to prevent drying. Use plastic film, glass, or moisture-retaining material for most foods except bread, breaded products, or crisp foods. You can leave these foods uncovered to prevent sogginess. If you use plastic film when heating foods to temperatures that result in a steam build-up, be sure to pierce the film before heating so that steam can escape.

Use waxed paper to loosely cover foods. It will not retain steam as does a glass cover or plastic film, and it will protect the oven from spatters. Paper towels will absorb some moisture and will also protect the oven from spatters.

Shape and Composition of Food

Food that is flat and thin heats faster than food that is chunky. The greatest amount of heating takes place within 3/4 of an inch of the surface. The interior of large food items, or dense foods, is heated by heat transferred from the outer food layer. The most uniform heating occurs in flat doughnut-shaped foods.

Heating several foods together

Heating times are critical when a combination of food is heated together. Foods of the same amounts and with similar moisture contents can be heated together successfully. Other foods may require some of the following adjustments:

1. Placement: place foods with the shortest heating times, such as low moisture, high fat, and high sugar foods, in the center of the plate, with the smallest or narrowest parts of the foods near the center.
2. Size and shape of portion: reduce the size of the portion, make two portions, or change the shape of the food, such as depressing the center of mashed potatoes.
3. Use of cover: allow foods to stand approximately 1 minute before removing the cover.
4. Initial temperature: allow more heating time for refrigerated foods than for foods of room temperature.
5. Additions: either add foods to the plate or put them in separate side dishes.

ADAPTING YOUR OWN RECIPES

When initially acquainting yourself with the Radarange Oven, it is best to follow recipes in this Cook Book that are similar to your own. Once you've become accustomed to the time factors in microwave cooking, you will be able to adapt many of those cherished, family favorite recipes to microwave cooking.

As a rule of thumb, when first starting to cook your recipes in the Radarange Oven, set the timer short of the time you estimate will be needed. You can always add a few seconds or minutes to finish the cooking job if the food is not quite done. When you are more experienced, we suggest taking some of your favorite recipes and changing them to 1/4 of the conventional cooking time.

5

Everyday foods – timed at a glance

Butter, melted	3/4 minute-1/2 cup
Butter, softened	1/2 minute-1/2 cup
Brownies	5 to 6 minutes/8 x 8 x 2-inch
Cakes	9 to 11 minutes/12 x 8 x 2-inch
Cakes, Coffee	7 to 8 minutes/10 x 6 x 2-inch
Cupcakes	1-1/2 to 2 minutes/6 cupcakes
Cookies, Individual	2-1/2 to 3 minutes/12 cookies
Chicken, Fryer	21 minutes/2-1/2-lb. fryer
Chocolate, melted	1 to 1-1/2 minutes/2 squares
Eggs	1/2 to 3/4 minute/egg
Fish	4 minutes /lb.
Chops	25 minutes/5 to 6 chops
Ham, whole	6 minutes/lb.
Ham, sliced	7 minutes/lb.
Meat Loaf	14 to 16 minutes/1-1/2 lb. loaf
Meatballs	6 to 7 minutes/lb.
Meat Patties	7 to 8 minutes/lb. - well done
Milk, hot	2 to 3 minutes/cup
Poultry, Roast	7 minutes/lb.
Potatoes	2-1/2 to 3 minutes/average 4 oz. potato
Pudding	5 to 6 minutes/4 servings
Pies, pastry shell	3 minutes/9-inch shell
Pies, 2-crust fruit	7 to 8 minutes/9-inch pie
Rolls, hot	1/4 to 1/2 minute/2 rolls
Roasts, Beef or Pork	7-9 minutes/lb.
Sandwiches	1/2 minute/sandwich
Sauces	2 to 3 minutes/cup
Soup, hot	1-1/2 to 2 minutes/cup
Vegetables, canned	1 to 2 minutes/cup
Vegetables, fresh	6 to 9 minutes/4 servings
Vegetables, frozen	5 to 7 minutes/10-oz. (each) pkg.
Vegetables, frozen	10 to 13 minutes/two 10-oz. (each) pkgs.
Water, hot	1 minute/cup
Water, boiling	2 to 3 minutes/cup
Wiener, in a bun	20 seconds/wiener

6
Convenience Foods

FOODS	YIELD OR SERVINGS	SUGGESTED COOKING UTENSIL	RADARANGE TIMING	SPECIAL INSTRUCTIONS
MAIN COURSES				
Cheeseburger Macaroni Cass.	5 cup serving	10" ceramic skillet	8 minutes	Stir after 4 minutes
Chicken Chow Mein	1 serving	8-9 inch plate	2-1/2—3-1/2 min.	Turn dish after every minute
Chow Mein	14 oz.	Ceramic casserole	2-1/2—3 minutes	Stir after 1-1/2 minutes
Fried Chicken	1 lb.—6 pieces	Serving plate	6 minutes	Turn dish after 3 minutes
Hamburger Helper Casserole	5 cup serving	10" ceramic skillet	8 minutes	Stir after 4 minutes
Lasagna	1 lb. 5 oz.	1-1/2-quart casserole	8 minutes	Turn dish every 2 minutes
Spaghetti & Meatballs	8 oz.	Sauce dish	2-1/2—3 minutes	Stir after 1-1/2 minutes
Stuffed Cabbage Rolls	7 oz.	1-quart casserole	8 minutes	Turn rolls every 2 minutes
Stuffed Green Peppers	7 oz.	1-quart casserole	6 minutes	Re-position peppers every 2 minutes
T.V. Dinners	1 serving	Glass dinner plate	4-5 minutes	Turn plate every 2 minutes
FROZEN DAIRY & CREAM SAUCES				
Creamed Chipped Beef	11 oz.	1 pint casserole	4-1/2 minutes	Stir every 2 minutes
Hungarian Cauliflower	12 oz.	1 pint casserole	6 minutes	Stir every 3 minutes
Macaroni & Cheese	8 oz.	Sauce dish	2-1/2—3 minutes	Stir after 1-1/2 minutes
Potatoes au Gratin	1-1/2 servings	1 quart casserole	4-1/2 minutes	Stir after 2 minutes
Scalloped Corn	12 oz.	1 pint casserole	4-1/2 minutes	Stir after 2 minutes
Scalloped Potato Casserole	12 oz.	1 quart casserole	3 minutes	Stir after 1-1/2 minutes
SOUPS N' STEWS				
Beef Stew	8 oz.	Soup bowl	3—4 minutes	Stir after 2 minutes
Chili	8 oz.	Soup bowl	3—4 minutes	Stir after 2 minutes
Pork Chop Suey	8 oz.	Soup bowl	3—4 minutes	Stir after 2 minutes
Vegetable Soup	8 oz.	Soup bowl	3—4 minutes	Stir after 2 minutes

7
Metric Conversion Table

Metric Equivalents		
1 milliliter	=	.001 liters
1 gram	=	.001 kilogram
Liquid Measure		
1 teaspoon	=	4.9 milliliters
1 tablespoon	=	14.8 milliliters
1 ounce	=	29.6 milliliters
1/4 cup	=	59.1 milliliters
1/3 cup	=	78.9 milliliters
1/2 cup	=	118.3 milliliters (1/8 liter)
1 cup	=	236.6 milliliters (1/4 liter)
1 pint	=	473.2 milliliters
1 quart	=	.9 liters
1 gallon	=	3.8 liters
1 milliliter	=	.2 teaspoons
	=	.07 tablespoon
	=	.004 cup
	=	.03 ounce
1 liter	=	1.1 quarts
	=	.26 gallons
Dry Measure		
1 ounce	=	28.6 grams
1 gram	=	.04 ounce
	=	.002 pounds
1 pound	=	454 grams

9 Meal Planning

Once you cook a few foods in the Radarange Oven, you will soon want to combine items to make a meal. Usually, meal preparation requires cooking items in sequence, one after another. Several items can be heated or cooked in the oven together, but this heating is usually not even, since one food may attract more microwave energy than the other.

Cooking in sequence requires only a little planning so that all foods are piping hot and ready for the table simultaneously. While planning a cooking sequence, consider the following:

What is the desired serving temperature?
What is the cooking time of each food?
How long will the food remain hot?
Does the food reheat easily?
Can the cooking time be broken into segments?

If an item needs cooling or chilling before serving, it should be prepared first. For items that need to be hot at the same time, start with either the largest item which will hold heat well, or those items that have more than one cooking step. Then, while one food is cooking, prepare the other foods to be cooked. If you have a meal consisting of quick-cooking foods, you may even find it easier to set the table before starting to cook.

Items which are often cooked first include: large cuts of meat or poultry, meat items that require simmering, casserole mixtures, and foods that are partially cooked before the final cooking or heating period. When a meat or casserole requires a long cooking period, the time can often be divided into 2 or 3 shorter periods. Often this improves the cooking of the meat or casserole, and makes the total meal preparation easier.

Items that are usually cooked closer to serving time include: fish items, small meat items that cook quickly, eggs, vegetables and sauces.

Items that are always cooked last include breads as well as foods that are cooked earlier but need some reheating before serving.

Here, for example, is a suggestion for preparing a dinner that includes meat loaf, baked potatoes, a vegetable and rolls. Start with the meat loaf because it requires the most time and it will hold heat the longest. However, since the potatoes also hold heat well, you may wish to cook the meat loaf about 3/4 done and let it stand while the potatoes cook. Then, while the potatoes are standing, complete cooking the meat loaf. While both the potatoes and meat loaf are standing, cook the vegetable. Finally, as the food is being placed on the table, heat the rolls.

Or, try this way of fixing a quick lunch consisting of a bowl of soup, hot dog and a cup of coffee. Heat the soup first since it takes the longest time and holds heat. Next, heat the coffee, and then heat the hot dog. If the soup has cooled too much, just return it to the oven for a minute to again bring it to a serving temperature.

The menus that follow on the next pages show how to apply these ideas in meal preparation. When convenient, you may wish to prepare part of the foods ahead so that at meal time the Radarange Oven is used primarily to heat the foods to a serving temperature. This is especially nice when you are entertaining and you wish to keep the last-minute kitchen time to a minimum. It is also convenient if you have a large family and need to cook in large quantities.

Complete meals can be easily prepared in the Radarange Oven. Start with simple meals and think through the sequence. Always remember that no matter what unexpected interruption develops, the foods will reheat beautifully. Begin with the ideas on the next pages. Then, go on to combine your own meal time favorites.

Duckling Bordeaux

10
Meal Planning

FAMILY BREAKFAST MENU	MENU TIME GUIDE
Orange Juice Bacon Scrambled Eggs Pancakes Coffee Milk	About 20 minutes before serving: 1. Heat syrup. 2. Soften butter. 3. Cook bacon while mixing pancakes. 4. Cook eggs while cooking pancakes on griddle. 5. Just before serving: Prepare coffee. 6. Reheat bacon 1 minute in Radarange Oven. 7. Reheat pancakes, if necessary, in Radarange Oven.

Bacon

8 strips bacon

1. Arrange bacon on rack in 2-quart utility dish. Cover with paper towel.
2. Cook in Radarange Oven 8 minutes, or until crisp.

Scrambled Eggs

3 tablespoons butter
8 eggs
1/2 cup milk
1 teaspoon chopped chives
1/2 teaspoon salt
1/4 teaspoon dry mustard

1. Heat butter in 1-quart casserole in Radarange Oven 1 minute, or until melted. Add remaining ingredients, beating until eggs are scrambled.
2. Cook, covered, in Radarange Oven 4-1/2 minutes, or until eggs are just about set. Stir twice during last half of cooking time.

   MICRO-TIP: When you anticipate needing to reheat eggs, undercook them slightly. Then they will finish cooking and reheat at the same time.

Hot Syrup

2 cups pancake syrup

1. Remove cap from syrup bottle. Leave syrup in bottle unless it is too tall for oven. If so, pour syrup into shorter pitcher before warming.
2. Heat in Radarange Oven 3 minutes, or until hot.

Softened Butter

1/2 cup butter (1 stick)

1. Unwrap butter. Place on serving plate.
2. Heat in Radarange Oven, using Automatic Defrost Control 45 seconds, or until softened.

   MICRO-TIP: When heating in Radarange Oven using Slo Cook or Automatic Defrost Control, decrease time to 30 seconds.

11
Meal Planning

BRUNCH FOR EIGHT MENU

Fresh Fruit Bowl
Crab and Egg Special
Hot Rolls
Coffee

MENU TIME GUIDE

Night before:
1. Prepare fruits.
2. Prepare crab dish through step 4.

About 15 to 20 minutes before serving:

3. Prepare coffee.
4. Complete crab dish, step 5.
5. Complete fruit bowl.
6. Heat rolls 2 to 3 minutes.

Crab and Egg Special

Although this recipe appears long, it is simple to serve because it can all be prepared ahead.

1-1/2 cups long-grain rice
3 cups water
1 teaspoon salt
1/2 teaspoon curry powder
2 tablespoons butter

1. Combine rice, water, salt, curry and butter in 2-quart casserole. Cook, covered, in Radarange Oven 8 minutes or until boiling. Then, cook in Radarange Oven, using Slo Cook or Automatic Defrost Cycle, for 13 minutes, or until rice is just about tender. Spread in 2-quart utility dish. Set aside.

8 hard-cooked eggs
1/4 cup mayonnaise
1 teaspoon lemon juice
1 teaspoon prepared mustard
1/2 teaspoon salt
Dash pepper
Paprika

2. Hard-cook eggs conventionally. Cool. Peel. Cut in half lengthwise and remove yolks to bowl. Mash well. Mix in mayonnaise, lemon juice, mustard, salt and pepper. Spoon mixture into whites. Sprinkle with paprika. Set aside.

3 tablespoons butter
2 cups sliced fresh mushrooms
3 tablespoons all-purpose flour
1/2 teaspoon salt
1-1/2 cups milk
1 tablespoon lemon juice

3. Combine butter and mushrooms in 2-quart casserole. Cook in Radarange Oven 4 minutes. Stir in flour, salt and milk. Cook in Radarange Oven 4 minutes, or until mixture boils, stirring several times. Stir in lemon juice.

1 (6 oz.) pkg. frozen crabmeat

4. Defrost crabmeat in Radarange Oven 1-1/2 minutes, or until just about defrosted. Add to sauce. Pour completed sauce over rice mixture. Mix lightly.

5. Cook rice mixture, covered with plastic wrap, in Radarange Oven 10 minutes, or until steaming hot. Stir once or twice during cooking. Arrange stuffed eggs on top of rice, making indentations in rice for each with spoon. Cover with plastic wrap. Heat in Radarange Oven 2 minutes, or until eggs are hot.

MICRO-TIP: When preparing ahead, store the rice mixture and the eggs in the refrigerator.

12
Meal Planning

LADIES LUNCHEON FOR SIX	MENU TIME GUIDE
Country Club Turkey Orange and Avocado Slices on Lettuce Perpetual Muffins Meringue-Topped Pot de Creme	Several hours ahead: 1. Prepare Pot de Creme, chill. 2. Prepare turkey through step 2. 3. Prepare and bake muffins in Radarange Oven. 4. Prepare orange and avocado slices. About 15 minutes before serving: 5. Complete turkey, step 3. 6. Assemble salad. 7. Heat muffins 2 to 3 minutes.

Country Club Turkey

2 (10 oz. each) pkgs. frozen asparagus spears

1. Cook frozen asparagus spears according to directions in Vegetable Section. Drain any liquid from asparagus and arrange spears evenly in buttered 8 x 8 x 2-inch glass baking dish.

1 (10-3/4 oz.) can cream of chicken soup
2 cups cubed, cooked turkey
1/2 cup sliced pitted ripe olives
1/2 cup diced pimiento
1/2 teaspoon grated onion
1/8 teaspoon nutmeg

2. Combine soup, turkey, olives, pimiento, onion and nutmeg. Spoon this mixture over asparagus.

2 tablespoons shredded smokey cheese

3. Top with cheese. Cook in Radarange Oven 8 minutes, or until heated through.

Meringue-Topped Pot de Creme

1-1/4 cups half and half
1 (6 oz.) pkg. semi-sweet chocolate morsels

1. Combine half and half and chocolate in 4-cup measuring cup. Heat in Radarange Oven 4 minutes, or until steaming hot. Beat until chocolate is completely blended.

1/4 cup sugar
1/4 teaspoon salt
3 egg yolks
1 teaspoon vanilla

2. Beat in sugar, salt, egg yolks and vanilla. Pour about 1/3 cup of mixture into each of six, 6-oz. custard cups. Cook in Radarange Oven, using Slo Cook or Automatic Defrost Cycle 4 minutes, or until mixture just begins to set.

3 egg whites
1/4 teaspoon cream of tartar
1/3 cup sugar
1/2 teaspoon vanilla

3. Beat egg whites and cream of tartar until frothy. Gradually add sugar, beating until soft peaks form. Beat in vanilla. Spoon onto chocolate mixture, forming peaks and swirls.

4. Cook in Radarange Oven using Slo Cook or Automatic Defrost Cycle 2-1/2 minutes, or until meringues are set. Cool. Refrigerate until served. If desired, garnish with twist of orange peel.

13
Meal Planning

FAMILY DINNER FOR FIVE	MENU TIME GUIDE
Chicken and Zucchini Delish Buttered Noodles Corn-in-the-Husk Simply Delicious Rice Pudding	About 40 minutes before serving: 1. Prepare pudding. 2. Prepare corn through step 1. 3. Prepare chicken through step 2. 4. Complete corn step 2 while cooking noodles on burner. 5. Complete chicken, step 3.

Chicken and Zucchini Delish

2 whole chicken breasts

1. Cut breasts in half. Remove and discard skin and bone. Cut chicken into bite-size pieces. Place in 2-quart casserole.

6 cups sliced zucchini
1 small sliced onion
1 minced clove garlic
2 tablespoons all-purpose flour
2 teaspoons instant chicken bouillon
1/2 teaspoon salt
1/4 teaspoon leaf thyme

2. Add zucchini, onion, garlic, flour, bouillon, salt and thyme. Mix well. Cook, covered, in Radarange Oven 11 minutes, or until just about tender. Stir mixture once or twice during cooking.

2 medium tomatoes
1 cup seasoned salad croutons

3. Wedge tomatoes and add to mixture. Cook, covered, in Radarange Oven about 2 minutes, or until tomatoes are heated through. Garnish with croutons.

MICRO-TIP: Cubed, cooked chicken or turkey can be substituted for chicken breasts. Use about 3 cups. The skin and bones can be boiled to prepare soup stock.

Simply Delicious Rice Pudding

1 cup instant rice
3 cups milk
1/4 teaspoon salt
1 (3-3/4 oz.) pkg. vanilla pudding mix

1. Combine rice, milk, salt and pudding in 1-1/2-quart casserole.

2. Cook in Radarange Oven 9 minutes, or until mixture boils. Stir every 2 minutes during last half of cooking time.

Nutmeg

3. Sprinkle with nutmeg. Cool. Serve warm or chilled.

Corn-in-the-Husk

5 ears corn

Remove outer husks, but leave inner husk on corn. Carefully remove silk. Replace husks and fasten with string or rubber band. Cook ears in Radarange Oven 9 minutes, or until steaming hot. Rearrange ears about halfway through cooking time. Serve with melted butter.

15
Meal Planning

MEXICAN SUPPER FOR SIX	MENU TIME GUIDE
Iced Fresh Tomato Soup **Mexitaco Pie** **Tossed Salad** **Bread Sticks** **Mexican Parfaits**	Early in day: 1. Prepare soup and chill. (See recipe p. 116) 2. Prepare greens for salad. 3. Prepare parfaits through step 1. Several hours ahead: 4. Complete parfaits, step 2. 5. Prepare Mexitaco Pie through step 1. About 10 minutes before serving: 6. Complete Mexitaco Pie, steps 2-3. 7. Complete salad.

Mexitaco Pie

1 lb. ground beef **3 cups chili con carne** **1 tablespoon instant minced green pepper** **1 tablespoon parsley flakes**	1. Brown beef in 1-1/2-quart casserole in Radarange Oven 5 minutes. Drain. Stir in chili con carne, green pepper and parsley flakes. Cook in Radarange Oven 5 minutes, turning dish halfway through cooking time. Set aside.
5-1/2 oz. pkg. tortilla chips	2. Crush half package of chips. Spread evenly in 9-inch pie plate. Use remaining whole chips to edge plate.
1 cup shredded cheddar cheese **1 tablespoon minced onion**	3. Sprinkle 1/2 cup cheese and 1-1/2 teaspoons minced onion over chips. Pour chili mixture into pie shell and top with remaining cheese and onion. Cook in Radarange Oven 2-1/2 minutes, or until cheese is melted.

Mexican Parfaits

1 (6 oz.) pkg. semi-sweet chocolate morsels **1/4 cup sugar** **2/3 cup evaporated milk** **1/4 teaspoon cinnamon, if desired** **1/2 teaspoon vanilla**	1. Combine chocolate, sugar and milk in 4-cup measuring cup. Cook in Radarange Oven 4-1/2 minutes, or until mixture boils and thickens. Stir every minute. Stir in cinnamon and vanilla. Cool. Refrigerate.
1 cup whipping cream **1/4 cup sugar** **1/2 teaspoon vanilla** **1 quart chocolate or coffee ice cream**	2. Whip cream until thickened. Beat in sugar and vanilla. Layer ice cream, chocolate sauce and whipped cream in 6 parfait glasses. Cover. Freeze until served. If desired, garnish with chopped nuts or chocolate curls.

Corn on the Cob

16
Meal Planning

CANDLELIGHT DINNER FOR FOUR	MENU TIME GUIDE
Duckling Bordeaux Parsleyed Rice Mandarin Carrots Gelatin Salad Orange Sherbet	Early in day: 1. Cook rice. (See recipe on page 151) 2. Prepare salad. 3. Defrost duckling if necessary. Several hours ahead: 4. Prepare duckling through step 4. 5. Prepare carrots through step 1. (See recipe on page 130) About 30 minutes before serving: 6. Complete duckling, steps 5-7. 7. Unmold salad. 8. Heat rice about 5 minutes. 9. Complete carrots, steps 2-3.

Duckling Bordeaux

4-5 lb. duckling

1. Preheat Amana Browning Skillet in Radarange Oven 4-1/2 minutes.
2. Quarter duckling. Place two pieces of duckling skin-side-down in skillet. Cook in Radarange Oven 5 minutes, turning duckling over halfway through cooking time. Drain.
3. Preheat Amana Browning Skillet again in Radarange Oven 2 minutes. Repeat step 2.

1/2 cup marmalade
1 tablespoon soy sauce

4. Mix marmalade with soy sauce. Reserve 1/4 cup mixture. Use remainder to baste duckling.
5. Arrange all duckling quarters in skillet and cook in Radarange Oven 20 minutes, turning dish every 5 minutes.
6. Cover, and let stand while preparing sauce.

1 tablespoon butter
1 tablespoon all-purpose flour
3/4 cup white wine
1/3 cup chicken broth
1 tablespoon vinegar
1/4 teaspoon pepper

7. Melt butter. Stir in flour and then remaining ingredients. Cook in Radarange Oven 1 minute. Add remaining marmalade. Cook an additional 1-1/2 minutes until smooth. Spoon over duckling.

Meal Planning

HOLIDAY DINNER FOR TEN	MENU TIME GUIDE
Roast Capon Dressing Sweet Potatoes in Orange Shells Buttered Broccoli Pumpkin Pie	Early in day: 1. Prepare pies. (See recipe on page 201) 2. Prepare potatoes through step 2. (See recipe on page 134) 3. Defrost capon if necessary. 4. Prepare capon through step 2. About 1-1/4 hours before serving: 5. Complete capon, steps 3-5. 6. Complete potatoes, step 3. 7. Cook broccoli using times in Vegetable Chapter. (See recipe on page 140)

Roast Capon

1/2 cup butter
1 (4 oz.) can drained mushrooms
2 stalks chopped celery
1/2 cup chopped onion

1. Combine butter, mushrooms, celery and onion in mixing bowl. Cook in Radarange Oven 7 minutes, or until vegetables are barely tender. Stir halfway through cooking time.

10 cubed slices bread
1 teaspoon salt
1 teaspoon poultry seasoning
1 teaspoon Worcestershire sauce

2. Stir in bread, salt, poultry seasoning and Worcestershire sauce.

7 to 8-lb. capon
Salt

3. Sprinkle cavity of capon lightly with salt. Spoon stuffing mixture into cavity. Truss openings with thread.

2 tablespoons vegetable oil
1/2 teaspoon paprika
1 teaspoon leaf tarragon
1 teaspoon soy sauce

4. Combine oil, paprika, tarragon and soy sauce. Brush over capon.

5. Place on cooking grill in 2-quart utility dish. Cook in Radarange Oven as follows:

First side up:	14 minutes
Back up:	14 minutes
Second side up:	14 minutes
Breast up:	14 minutes

Cook about 8 minutes per pound, or until meat thermometer inserted in breast registers 165° .

19
Cooking for One or Two

Good meals for one or two need no longer be a problem. With your Radarange Oven, you can cook in small quantities to fit any need. Or, cook larger quantities and refrigerate or freeze the extras. These can easily be reheated for other meals.

Although this chapter is called cooking for one or two, it is also a source of ideas for flexible cooking. Interesting meals with a minimum of preparation require only a little pre-planning. Prepare a larger quantity of a basic item, but instead of having the same old left-overs for several days, have "planned-overs". Just vary the seasoning, sauce, shape or accompaniments so that each day makes it an interesting new meal with a minimum of preparation. Use the freezer, too, to add even more flexibility to this idea of "planned-overs".

So, whether your meal time means cooking for one, two or several, use this chapter as a source for interesting, flexible meal time ideas as well as for cooking small quantities of food.

Slice and Heat Cheese Spread

About 40 snacks to slice and heat when desired.

1-1/2 ozs. cream cheese
1 cup shredded Cheddar cheese
1 tablespoon all-purpose flour
1 tablespoon minced onion
1 teaspoon lemon juice
1/2 teaspoon prepared mustard
1/4 teaspoon Worcestershire sauce

1. Heat cream cheese in glass bowl in Radarange Oven 15 seconds, or until softened. Stir in remaining ingredients except crackers. Mix well. Place mixture on waxed paper, forming roll about 11 inches long and 1 inch thick. Refrigerate or freeze. If frozen, allow to thaw before slicing.

Crackers

2. Slice cheese roll into 1/4-inch thick slices. Place cheese slices on crackers. Arrange 12 crackers at a time on glass tray.
3. Heat in Radarange Oven 1 minute, or until cheese is melted. Repeat process with remaining cheese and crackers.

Spicy Fruit Punch Mix

A mix for 8 servings, to use as needed.

1 (3 oz.) pkg. sweetened lemonade mix
1 (3 oz.) pkg. sweetened orange drink mix
3 cinnamon sticks, broken into pieces
1/2 teaspoon whole cloves
1 teaspoon whole allspice

1. Combine all ingredients.
2. Heat water in glass mugs or cups in Radarange Oven about 1-1/2 minutes per cup, or until steaming hot. Stir in 2 rounded teaspoons of mix per cup.

Little Meat Loaves

20
Cooking for One or Two

Basic Meat Balls

18 meat balls to finish with a sweet-sour sauce or Mornay sauce (see recipes below).

1 lb. ground beef
1 small chopped onion
1/3 cup dry bread crumbs
1/4 cup milk
1 egg
3/4 teaspoon salt
1/8 teaspoon pepper

1. Combine all ingredients in mixing bowl. Mix well. Form into about 18 meat balls, 1-1/2 inches in diameter. Arrange in 8-inch round glass baking dish.
2. Cover with waxed paper. Cook in Radarange Oven 6 to 7 minutes. Turn halfway through cooking time. Drain.

   MICRO-TIP: Serve with one of the sauces below.

Sweet-Sour Meat Ball Sauce

Just right for 2 or 3 servings.

1 (8-1/4 oz.) can pineapple chunks
1/2 tablespoon cornstarch
3 tablespoons brown sugar
2 tablespoons vinegar
1 tablespoon soy sauce

1/2 chopped green pepper

1. Drain pineapple juice into 1-quart glass casserole. Stir in cornstarch, brown sugar, vinegar and soy sauce.
2. Cook in Radarange Oven 1 minute until mixture boils. Stir halfway through cooking time. Stir in green pepper and pineapple.
3. Cook, covered, in Radarange Oven 3 minutes. Stir halfway through cooking time. Pour over meat balls.

Mornay Sauce

Even though the meat balls are the same, this sauce adds a totally new taste: 2 to 3 servings.

2 tablespoons butter
2 tablespoons all-purpose flour
1 teaspoon instant chicken bouillon
Dash pepper
1-1/4 cups milk

1/4 cup shredded Swiss cheese
2 tablespoons grated Parmesan cheese
1 teaspoon dry parsley flakes

1. Heat butter in 1-quart glass casserole in Radarange Oven 45 seconds or until melted. Blend in flour, bouillon, pepper and milk.
2. Cook in Radarange Oven 4 minutes, or until mixture boils, stirring occasionally. Stir in Swiss cheese. Sprinkle with Parmesan cheese and parsley.
3. Cook, covered, in Radarange Oven 4 minutes, or until thoroughly heated.

   MICRO-TIP: Use shorter heating time when meat balls are freshly cooked, and longer heating time when they have been refrigerated.

21
Cooking for One or Two

Fish for One

Start with either fresh or frozen fish and cook it right on your dinner plate.

1/4 lb. fish fillet
1/2 tablespoon butter
1 teaspoon lemon juice
Salt
Pepper

1. Arrange fish on glass serving plate. Dot with butter. Sprinkle with lemon juice, salt and pepper.
2. Cover with waxed paper. Cook in Radarange Oven about 1-1/2 minutes for fresh fish, and 3 minutes for frozen fish. Fish should flake apart easily. Drain excess liquid from plate before serving.

Little Meat Loaves

4 little meat loaves for four hearty appetites.

1-1/2 lbs. ground beef
1 cup bread crumbs
3/4 cup evaporated milk
1 egg
1 tablespoon instant minced onion
1-1/2 teaspoons salt
1/2 teaspoon Italian seasoning
1/4 teaspoon seasoned pepper

1. Combine first eight ingredients in large bowl. Mix well. Shape into 4 individual loaves. Place loaves in 2-quart glass utility dish.
2. Cook in Radarange Oven 10 minutes, turning dish at 3-minute intervals. Turn loaves over halfway through cooking time.

1/3 cup orange marmalade
1 teaspoon vinegar
1/2 teaspoon gravy mix

3. Mix glaze ingredients. Brush or spoon over tops and sides of loaves. Cook in Radarange Oven 2 minutes, turning halfway through cooking time. Let stand 5 minutes before serving.

MICRO-TIP: For two meat loaves, arrange in 8 x 8 x 2-inch glass baking dish. Cook in Radarange Oven 6 minutes. Turn loaves, brushing top and sides with glaze. Cook in Radarange Oven 2 minutes, or until done.

Turkey and Dressing for Two

Here is a way to enjoy turkey and dressing without so many leftovers.

1 hind quarter turkey (3-1/4 to 3-3/4 lbs.)
Salt
Pepper
1/2 cup water

1. Sprinkle inside part of thawed turkey with salt and pepper. Place in 2-quart glass casserole including extra turkey parts. Place extras under larger piece. Add water.

2 cups crumbled herb-seasoned stuffing mix
2 tablespoons melted butter

2. Cook, covered, in Radarange Oven 20 minutes. Turn turkey over halfway through cooking time. Remove turkey from broth. Stir dressing mix into broth. Arrange turkey on dressing, placing skin-side-up. Brush with butter.
3. Cook in Radarange Oven, using Slo Cook or Automatic Defrost Cycle 20 to 25 minutes, or until meat is done.

22

Cooking for One or Two

Eggs Benedict for Two

Elegant egg dish for 2 to enjoy.

1 cup water
4 eggs

1. Heat water in covered 1-quart glass casserole in Radarange Oven 2 minutes until steaming hot. Break eggs into water.
2. Cook, covered, in Radarange Oven 2 to 3 minutes until prepared as desired. Let stand, covered.

4 thin slices Canadian-style bacon or ham
2 tablespoons butter
1 egg yolk
1 teaspoon lemon juice
1/8 teaspoon salt
1/8 teaspoon dry mustard

3. Heat Canadian bacon on plate in Radarange Oven 45 seconds. Cover with paper towel.
4. Heat butter in small dish in Radarange Oven 45 seconds until melted. Beat egg yolk with fork in 1-cup glass measuring cup. Beat in butter, lemon juice, salt and mustard.
5. Cook in Radarange Oven 20 seconds. Beat with fork. To thicken slightly, cook 10 seconds more.

2 split English muffins

6. Arrange each English muffin, cut-side-up, on serving plate. Top each half with slice of bacon. Remove egg from water, using slotted spatula or spoon. Place egg on top of bacon. Spoon sauce over eggs.
7. Heat one plate at a time in Radarange Oven 15 seconds.

Bacon and Eggs for One

Cook bacon and eggs together in an Amana Browning Skillet.

2 slices bacon

1. Preheat 6-1/2-inch Amana Browning Skillet in Radarange Oven 2 minutes. Cut bacon in half and arrange in skillet.

1 egg
Salt
Pepper

2. Cover with paper towel and cook in Radarange Oven 1 minute. Turn bacon over and arrange at edge of skillet. Break egg into center of skillet.
3. Cook, covered, in Radarange Oven about 1 minute, or until prepared as desired. Season egg.

MICRO-TIP: If cooking 2 eggs, increase time in step 3 to 1-1/4 or 1-1/2 minutes.

Frozen Vegetables for One

No fuss—
No waste.

1/2 cup (1/3 of 10-oz. pkg.) frozen vegetable

Place vegetable in small glass serving dish. Cook, covered, in Radarange Oven 2-1/2 minutes. Drain if necessary. Season to taste.

Carrots for Two

Cooked carrots in a small quantity.

1 cup sliced carrots
1 tablespoon water
Salt

Combine carrots and water in small glass serving dish. Cook, covered, using Slo Cook or Automatic Defrost Cycle in Radarange Oven 6-1/2 minutes, or until tender. Salt as desired.

23
Cooking for One or Two

Quick Scalloped Potatoes

2 easy servings that start with frozen hash browns.

6 ozs. frozen hash browns

1 teaspoon instant minced onion
1 tablespoon butter
1/2 cup milk
1/4 teaspoon salt
Dash pepper
Paprika

1. Place frozen potatoes in 1-quart casserole.
2. Cook, covered, in Radarange Oven 4 minutes. Mix in onion, butter, milk, salt and pepper. Sprinkle with paprika.
3. Cook, covered, in Radarange Oven 4 to 5 minutes, or until liquid is absorbed and potatoes are tender.

Potato Mix for Two

If entire package is too much, just prepare enough for 2.

1 cup boiling water

1/3 cup milk
1/2 (5-1/2 oz. size) pkg. au gratin potato mix
1 tablespoon butter

1. Pour water into 1-quart glass casserole.
2. Stir in milk, potatoes, sauce mix and butter. Cook in Radarange Oven 2 minutes, or until mixture boils.
3. Using Slo Cook or Automatic Defrost Cycle, cook in Radarange Oven 8 to 9 minutes, or until potatoes are tender.

Store-and-Bake-Bran Muffins

Better batter for 14 muffins.

1-1/2 cups bran cereal
1 cup warm water
1 cup buttermilk or sour milk
1/4 cup oil
1 egg
1-1/4 cups all-purpose flour
3/4 cup raisins or chopped dates
1/2 cup firmly packed brown sugar
1 teaspoon salt
1 teaspoon soda

1. Combine bran cereal and water in mixing bowl. Blend in remaining ingredients. Prepare muffins immediately, or refrigerate batter until needed.
2. Lightly grease fourteen, 6-oz. glass custard cups. Pour 1/4 cup batter into each. Place 3 cups in circular arrangement in Radarange Oven. Cook 2-1/2 minutes. Repeat process with remaining muffins.

24
Cooking for One or Two

Quick Cobbler

2 servings for today — 2 for warming and serving tomorrow.

1 (16 oz.) can sliced peaches

1 cup biscuit mix
3 tablespoons sugar
1/4 teaspoon almond extract
1/3 cup milk
1 tablespoon butter
1/8 teaspoon cinnamon

1. Place peaches with juice in 8-inch round glass baking dish. Set aside.
2. Combine biscuit mix, 2 tablespoons sugar, extract and milk. Spoon over peaches. Cut butter into small pieces. Scatter over batter. Combine 1 tablespoon sugar and the cinnamon. Sprinkle over top.
3. Cook in Radarange Oven 5-1/2 minutes. Serve warm.

Individual Apple Crisps

Servings for 2 hungry appetites.

2 medium cooking apples
1 tablespoon water

3 tablespoons butter
1/4 cup all-purpose flour
1/4 cup quick rolled oats
1/3 cup firmly packed brown sugar
1/2 teaspoon cinnamon

1. Slice and peel apples into two 1-cup serving dish. Sprinkle with water. Set aside.
2. Heat butter in Radarange Oven 1/2 minute until softened. Mix in flour, oats, brown sugar and cinnamon until crumbly. Sprinkle over apples.
3. Cook in Radarange Oven 5-1/2 minutes, or until apples are tender.

Easy Apple Pie

Filling on the outside — crust on the inside — 3 deliciously different apple pies.

1/2 pie crust stick
1 tablespoon water

3 cups sliced cooking apples
1/2 cup sugar
1 tablespoon all-purpose flour
1 teaspoon lemon juice
1/2 teaspoon cinnamon
1 tablespoon butter

1. Prepare pastry with 1 tablespoon water as directed on package. Divide dough into thirds. Roll out to 4-inch circle. Arrange on paper towel.
2. Cook in Radarange Oven 3 minutes, or until crust has dry flaky appearance. Cool.
3. Combine remaining ingredients in 1-quart glass casserole.
4. Cook, covered, in Radarange Oven 5 minutes, or until apples are tender. Stir halfway through cooking time. Cool slightly. Spoon half of warm apples onto 3 serving plates. Top each with pastry circle and more apples.

25
Cooking for One or Two

Individual Caramel Custards

This recipe makes 4 custards—just enough for two meals.

4 tablespoons caramel ice cream topping

1 (3 oz.) pkg. cream cheese
3 tablespoons sugar
1/4 teaspoon vanilla
2 eggs
3/4 cup milk

1. Spoon 1 tablespoon ice cream topping into four, 6-oz. custard cups. Set aside.
2. Heat cream cheese in Radarange Oven 30 seconds until softened. Blend in sugar and vanilla. Beat in eggs one at a time. Mix in milk. Pour custard mixture over caramel topping.
3. Using Slo Cook or Automatic Defrost Cycle, cook in Radarange Oven 8 minutes, or until knife inserted in center comes out clean. Cool and refrigerate. Invert onto serving plates.

Flexible Cake Mix

You no longer need to serve the same cake day after day. Here are 2 ways to vary a one-layer mix — 4 orange cupcakes and one chocolate cream layer.

1 (9 oz.) pkg. yellow cake mix

1. Prepare cake mix batter as directed on package. Pour half of batter into four lightly greased 6-oz. glass custard cups, filling each half full. Pour remaining batter into lightly greased 8-inch round glass baking dish.
2. Cook cupcakes in Radarange Oven 2 minutes using Slo Cook or Automatic Defrost Cycle. Then cook in Radarange Oven 2 minutes until top is no longer doughy. Cook layer in same way.

1/4 cup orange juice
1/4 cup sugar
2 tablespoons butter
1 tablespoon orange-flavor liqueur
Sweetened whipped cream

For Orange Cupcakes, heat orange juice, sugar and butter in 1-cup glass measuring cup 1-1/2 minutes, until boiling. Stir in liqueur. Spoon over cupcakes in custard cups. Cool. Serve topped with whipped cream.

1/2 (3-3/4 oz.) pkg. chocolate pudding mix
1/2 tablespoon unflavored gelatin
1 cup milk
1/2 cup whipping cream

For Chocolate Cream Cake, combine pudding mix, gelatin and milk in 2-cup glass measuring cup. Cook in Radarange Oven 3-1/2 minutes, until mixture boils. Stir every minute. Cool. Beat cream until thickened. Fold in chocolate pudding. Spread over cake in baking dish. Cover, and refrigerate. Cut into wedges to serve.

Mix and Match Dessert Crunch

2 cups of crunch to make quick desserts from fruit or pudding.

1/2 cup butter or margarine
1-1/4 cups all-purpose flour
1/3 cup firmly packed brown sugar
1/2 teaspoon nutmeg
1/2 cup chopped nuts

1. Heat butter in 9-inch glass pie plate in Radarange Oven 3/4 minute until softened. Mix in remaining ingredients with fork until crumbly.
2. Cook in Radarange Oven 5-1/2 minutes until lightly toasted. Stir several times. Cool. Serve as topping on ice cream, pudding, fruit, pie filling or whipped cream.

27
Low Calorie Cooking

Controlling calories and cooking with microwaves are natural partners. Fat, butter and margarine are not needed in excess to prevent sticking. Sauces can be eliminated or kept to a minimum, and menus can be individualized to meet diet needs.

The quick "steam-type" cooking made possible with the Radarange Oven allows foods to steam in their natural juices. Just add a touch of a favorite seasoning to enhance natural food flavors. Or, try cooking that favorite food in a little bouillon, rather than using a tempting rich sauce or butter.

With the quick-cooking capabilities of your Radarange Oven, you can quickly prepare a nutritious type snack at any time of the day or night. The snack will be ready to eat before you have time to become tempted by any higher calorie snacks.

Another way to reduce calories is to eat less than a normal serving. With the easy-to-reheat feature of the Radarange Oven, a single portion can be enjoyed at one meal and any remainder can be saved for reheating at another meal.

Many recipes in this chapter will be enjoyable whether you are counting calories or not. When part of the family wishes to enjoy higher calorie sauces and seasonings, place the calorie controlled portion in a small dish inside the regular baking dish. This way you can keep the extra calories out of the small dish, and yet cook the food together without altering the cooking time.

We have presented a sampling of recipes here that are low in calories, yet help to make good nutritious meals. Use these as ideas for turning your other favorite recipes into low calorie specialties whenever the need arises.

Spring-Time Sole

Each of 4 servings contains 138 calories.

3/4 lb. fresh asparagus
1 tablespoon water

1 lb. sole fillets
1/2 teaspoon salt
1/4 teaspoon leaf thyme

1/2 cup plain yogurt
1 teaspoon buttermilk-type salad dressing mix

1. Remove tough ends from asparagus. Place asparagus in 10 x 6 x 2-inch glass baking dish. Add water.
2. Cover with plastic wrap. Pierce. Cook in Radarange Oven 6 to 7 minutes, until just tender.
3. Cut fillets into serving pieces. Sprinkle with salt and thyme. Divide asparagus among pieces, placing crosswise on fillets. Roll fillet around asparagus. Fasten with toothpick. Return each to baking dish.
4. Cover with plastic wrap. Pierce. Cook in Radarange Oven 4 to 4-1/2 minutes, until fish flakes easily with fork. Combine yogurt and dressing mix. Spoon over fillets.
5. Cook in Radarange Oven 1 minute until sauce is heated.

MICRO-TIP: With frozen asparagus, cook 5 to 6 minutes, or until just about done. With canned asparagus, omit steps 1 and 2.

Colorful Fillets

There are only 118 calories in each serving of this fish recipe that serves 4.

1 lb. haddock fillets, cut into serving pieces
3 sliced green onions
1 cup sliced fresh mushrooms
1 cup chopped tomato
3/4 teaspoon salt
Dash pepper
1/8 teaspoon leaf basil

Arrange fillets in 8 x 8 x 2-inch glass baking dish. Top with vegetables. Sprinkle with seasonings. Cover with plastic wrap. Pierce. Cook in Radarange Oven 4-1/2 minutes, until fish flakes easily with fork.

‹ *Grecian Eggplant*

28
Low Calorie Cooking

Beef Strips with Tomatoes

It is a delight to diet with a recipe like this that serves 5 and contains only 232 calories per serving.

1 (1 lb.) flank steak
1/3 cup soy sauce
1/3 cup dry white wine
1 teaspoon sugar

1. Cut steak across grain into thin strips. Place in 2-quart glass casserole. Combine soy sauce, wine and sugar. Pour over meat. Mix lightly to coat evenly. Allow to marinate 1 to 2 hours.

2 tablespoons cornstarch
1 medium sliced onion
2 cups sliced fresh mushrooms
1/2 sliced green pepper

2. Stir in cornstarch, onion, mushrooms and pepper.

1 pint cherry tomatoes

3. Cook, covered, in Radarange Oven 8 to 10 minutes, until sauce is thickened. Stir halfway through cooking time. Add tomatoes.
4. Cook, covered, in Radarange Oven about 1 minute until tomatoes are heated.

Crustless Pizza Pie

Five pizza-flavored servings with 210 calories each.

1 lb. extra lean ground beef
1/4 cup chopped onion
1/3 cup quick rolled oats
1 egg
1/2 teaspoon salt
1/8 teaspoon pepper
1 (8 oz.) can tomato sauce
1 teaspoon leaf oregano
Green pepper rings

1. Combine beef, onion, oats, egg, salt and pepper in mixing bowl. Combine tomato sauce and oregano. Mix half into meat mixture. Press into 9-inch glass pie plate. Arrange pepper rings in center.

1 (4 oz.) can drained mushroom stems and pieces
1/2 cup shredded Mozzarella cheese

2. Cover with waxed paper. Cook in Radarange Oven 10 minutes until meat is done. Turn dish every 3 minutes. Drain any excess juices. Spoon remaining tomato sauce over top. Sprinkle with mushrooms and cheese.
3. Cook in Radarange Oven 3 minutes until heated throughout.

Mint Julius Chicken

Four refreshing servings with only 160 calories each.

2 halved chicken breasts
1/3 cup thawed orange juice concentrate
1 tablespoon freshly chopped mint leaves
1/2 teaspoon salt

1. Arrange chicken breasts skin-side-up in 8 x 8 x 2-inch glass baking dish. Pour orange juice over chicken. Sprinkle with mint and salt.
2. Cover with waxed paper. Cook in Radarange Oven until chicken is done, 16 to 18 minutes.

Grecian Eggplant

Four low-calorie servings at 96 calories each.

1 medium eggplant

1. Cut eggplant in half lengthwise. Score surface crosswise, leaving shell intact. Remove and cube pulp of eggplant. Set aside.

2 slices bacon
1 cup chopped onion
1 large chopped green pepper
1 (1 lb.) can partially drained stewed tomatoes
1 teaspoon leaf oregano
1 teaspoon leaf basil
1/8 teaspoon minced garlic
1/8 teaspoon pepper

2. Cook bacon in 10-inch ceramic skillet in Radarange Oven 3 minutes. Remove bacon. Discard drippings. Add onion and green pepper to skillet. Cook in Radarange Oven 3 minutes. Stir halfway through cooking time. Add eggplant cubes, tomatoes and seasonings.

1 tablespoon chopped pimiento
1 tablespoon capers

3. Cook in Radarange Oven 8 to 10 minutes, or until eggplant is tender. Stir halfway through cooking time. Drain juices. Fill eggplant shells with vegetable mixture. Garnish with crumbled bacon, pimiento and capers.

MICRO-TIP: Grecian Eggplant can be prepared ahead of time, refrigerated, then reheated in Radarange Oven 3 minutes.

Oriental Pork

A colorful, tasty combination to serve 5. Just 173 calories per serving.

1 (1 lb.) pork tenderloin
1 tablespoon cornstarch
3 tablespoons soy sauce
1 teaspoon instant chicken bouillon

1. Remove and discard any fat from tenderloin. Cut into bite-size pieces. Combine pork, cornstarch, soy sauce and bouillon in 2-quart glass casserole.

4 cups thin-sliced Chinese cabbage
1 (6 oz.) pkg. frozen pea pods
1 (4 oz.) can drained bamboo shoots
1 (5 oz.) can drained, sliced water chestnuts

2. Cook, covered, in Radarange Oven 6 minutes, or until meat is no longer pink. Add vegetables.

3. Cook, covered, in Radarange Oven 6 to 7 minutes, or until meat and vegetables are cooked as desired. Stir once or twice during cooking.

Low Calorie Cooking

Snappy Beans

6 snappy servings with 20 calories each.

- **2 (10 oz. each) pkgs. frozen French-cut green beans**
- **1/4 cup plain yogurt**
- **1 teaspoon salt**
- **1/2 teaspoon prepared mustard**
- **1/4 teaspoon Worcestershire sauce**

Place beans in 1-1/2-quart glass casserole. Cook, covered, 10 minutes in Radarange Oven, until thawed and heated. Drain. Combine remaining ingredients. Mix lightly with beans. Cook in Radarange Oven 1 minute, or until heated through.

Onion Lovers' Corn

Corn-on-the-cob really can be good without butter. Just 182 calories per ear.

- **2 tablespoons dry onion soup mix**
- **1 tablespoon water**
- **4 ears frozen corn on the cob**

1. Combine soup and water in small dish. Spread on frozen corn. Place in 8 x 8 x 2-inch glass baking dish.
2. Cover with waxed paper. Cook in Radarange Oven 10 to 11 minutes until hot. Turn ears halfway through cooking time.

Spinach Salad

There are 6 nutritious servings of this salad with only 60 calories each.

- **10 oz. (8 cups) fresh spinach**
- **4 sliced green onions**
- **1 cup sliced mushrooms**
- **1 (5 oz.) can drained, sliced water chestnuts**

1. Wash spinach. Tear into bite-size pieces. Combine with onions, mushrooms and chestnuts in large bowl. Cover and refrigerate.

- **1/4 cup vinegar**
- **1 tablespoon sugar**
- **2 tablespoons water**
- **1/2 teaspoon salt**
- **1/4 teaspoon dry mustard**

2. Combine vinegar, sugar, water, salt and mustard. Heat in Radarange Oven 45 seconds, until boiling.

- **2 tablespoons imitation bacon bits**

3. Pour hot dressing over spinach mixture. Toss to mix evenly. Sprinkle with bacon bits.

Zucchini Boats

A nice combination to serve with sliced chicken or turkey. Each of the 8 halves contains about 64 calories.

- **4 small zucchini**
- **2 cups crumbled herb-seasoned stuffing mix**
- **1 tablespoon butter**
- **1/4 teaspoon salt**

1. Cut zucchini in half lengthwise. Scoop out center, using spoon, leaving 1/4-inch shell. Chop scooped-out pulp, and place in glass mixing bowl. Add stuffing mix, butter and salt. Mix lightly.

- **1 slightly-beaten egg**
- **2 tablespoons grated Parmesan cheese**

2. Cook, covered, in Radarange Oven 6 minutes, until stuffing is moistened. Stir in egg. Spoon into zucchini shells. Place on large serving platter. Sprinkle with Parmesan cheese.
3. Cover with waxed paper. Cook in Radarange Oven 8 minutes, until zucchini is tender.

MICRO-TIP: A grapefruit spoon is handy for scooping out zucchini.

31
Low Calorie Cooking

Slimline Apples

A way to enjoy cooked apple halves with only 35 calories per half.

4 medium baking apples
1 (12 oz.) can low-calorie strawberry-flavor soda beverage

1. Cut apples in half. Remove core. Place apples cut-side-down in 2-quart utility dish. Pour soda beverage over apples.
2. Cover with waxed paper. Cook in Radarange Oven 8 to 10 minutes, until apples are tender.

Strawberry Foam

Vary the fruit and gelatin flavor for these 5 servings. With strawberries, there are 71 calories per serving.

2 cups water
1 (3 oz.) pkg. strawberry-flavor gelatin

1. Heat 1 cup water in 4-cup glass measuring cup 2 minutes in Radarange Oven, until boiling. Add gelatin. Stir until dissolved. Blend in 1 cup cold water. Refrigerate until mixture starts to thicken.

1 egg white
1 cup sliced strawberries

2. Beat egg white at high speed until frothy. Gradually beat in thickened gelatin. Divide strawberries among five serving dishes. Pour gelatin mixture into dishes. Refrigerate until set, about 3 hours.

Light 'N' Fruity Tapioca

6 servings of light, creamy homemade pudding. Each serving contains about 100 calories.

2 cups skim milk
2 eggs
1/4 cup sugar
1/4 cup quick-cooking tapioca
1/4 teaspoon salt

1. Measure milk into 4-cup glass measuring cup. Separate eggs. Add 1/4 cup sugar, tapioca, salt and egg yolks to milk. Blend well. Cook in Radarange Oven 6 minutes until mixture boils. Stir after 4 minutes.

2 tablespoons sugar
1 teaspoon vanilla

2. Beat egg whites in large bowl until frothy. Gradually stir in sugar, beating until mixture forms soft peaks. Beat in vanilla. Fold in pudding mixture. Cool.

2 cups sliced fresh fruit

3. Spoon fruit into six serving dishes. Top with pudding.

33
Appetizers & Beverages

Your guests won't believe their eyes when they see the array of hot, flavorful canapes and hors d'oeuvres you "pop out" of the Radarange Oven. They will indeed think there is "magic" involved when they see Escargot, Oysters Casino and Clams prepared in seconds. At your next party try one of these simple but elegant recipes for:

Escargot

4 servings of snails from land or sea.

1/2 cup butter
1/2 teaspoon instant minced garlic
1 teaspoon fresh parsley flakes

1. In 1-cup measure, combine butter, garlic and parsley. Cook 1 minute or until butter bubbles.

1 (4-1/2 oz.) can snails
White wine
Snail shells

2. Place snails in compartments of 4 special 6-hole dishes or in 4 sauce dishes. Half-fill compartments with seasoned butter, or pour 1/4 of sauce into each dish. Refrigerate. Just before cooking pour 1 teaspoon of white wine over each snail. Place snail shells in glass pie plate. Cover with plastic wrap.
3. Cook in Radarange Oven 45 to 60 seconds, or until butter begins to bubble.

Coquilles

6 to 8 servings of scallops.

1 lb. fresh or frozen scallops

1. Defrost frozen scallops. Cut into fourths. Set aside.

1 lb. fresh mushrooms
5 tablespoons butter or margarine
2 tablespoons lemon juice

2. Wash, drain and slice mushrooms. Cook in 2 tablespoons butter and lemon juice in Radarange Oven 2 minutes. Stir halfway through cooking time. Drain.

1 cup dry white wine
1/4 teaspoon savory
1 bay leaf
1/2 teaspoon salt
1/8 teaspoon pepper

3. Combine wine, savory, bay leaf, salt and pepper in 1-1/2-quart glass dish. Add scallops. Cook in Radarange Oven 3 minutes. Drain. Reserve 1 cup of broth.

3 tablespoons all-purpose flour
1 cup light cream

4. Using 2-quart glass dish, melt 3 tablespoons butter and blend with flour to make smooth paste. Gradually stir in broth, then cream. Cook in Radarange Oven 5 to 5-1/2 minutes, or until sauce thickens. Stir well every 30 seconds.
5. Stir scallops and mushrooms into sauce. Cook in Radarange Oven 6 to 7 minutes, or until thoroughly heated.

1/2 cup toasted bread crumbs
Paprika

6. Spoon into seashells or ramekins to serve. Garnish with sprinkling of bread crumbs and paprika.

Oysters Casino, Escargot, Clams, Coquilles

Bacon-Cheese Puffs

60 bacon beauties.

1 lb. softened cream cheese
1 egg yolk
1 teaspoon instant minced onion

1. Combine cream cheese, egg yolk and onion in large mixer bowl. Beat until smooth and blended.

1 teaspoon baking powder
2-1/2 tablespoons imitation bacon pieces

2. Stir in baking powder and bacon pieces. Blend well.

60 round butter crackers

3. Place 6 crackers on paper plate in circle. Top each cracker with 1/2 teaspoon of cheese mixture. Heat in Radarange Oven for 30 seconds. Serve hot.

MICRO-TIP: Bacon-cheese mixture may be covered and stored in refrigerator until ready to use.

Hot Roquefort Canapes

Tantalizing tray of treats.

1 (3 oz.) pkg. cream cheese
1/4 cup crumbled Roquefort cheese
1/4 cup finely chopped pecans
1/2 teaspoon Worcestershire sauce
Dash Tabasco

1. Blend together first five ingredients.

Butter crackers

2. Spread mixture on crackers, using about 1 teaspoon for each cracker.
3. Arrange 12 on piece of waxed paper laid on paper towel in Radarange Oven. Heat for 50 to 60 seconds, until cheese spread begins to bubble. Serve warm.

Hot Crabmeat Canapes

25 canapes for that hungry cocktail crowd.

1 (6 oz.) pkg. frozen crabmeat
1 cup mayonnaise
1 teaspoon lemon juice

1. Thaw crabmeat. If crabmeat is in large pieces, shred with fork. Stir crabmeat into mayonnaise and season with lemon juice.

1 egg white

2. Beat egg white until stiff but not dry. Fold into crabmeat mixture.

Melba toast

3. Place approximately 1 teaspoon of crab mixture on 2-inch squares of Melba toast. Arrange on serving dish.
4. Cook in Radarange Oven 1-1/2 to 1-3/4 minutes for 12 canapes, or 3 to 3-1/2 minutes for 25 canapes. Serve immediately.

Appetizers and Beverages

Snack-Time Kabobs

60 snacks simple to prepare.

1 (12 oz.) can luncheon meat
1 (13-1/4 oz.) can pineapple chunks

Round toothpicks

1 tablespoon brown sugar
2 tablespoons soy sauce
1 tablespoon vinegar

1. Cut meat into about 60 small cubes. Drain pineapple chunks and cut each chunk in half.
2. For each kabob, string 1 chunk of meat and 1 piece of pineapple on toothpick. Arrange kabobs one layer deep in shallow dish.
3. Combine remaining ingredients and spoon over kabobs. Refrigerate for 1 hour. Transfer to larger plate.
4. Cook approximately 30 or half of kabobs in Radarange Oven about 2 minutes. Turn kabobs, spoon sauce over them, and cook additional 1-3/4 minutes. Repeat with other half of kabobs when ready to serve.

Shrimp Piquant

15 to 18 spicy shrimp hors d'oeuvres.

1 lb. fresh shrimp in shells

1/4 cup butter or margarine
3/4 teaspoon chili powder
Dash garlic powder

2-3 slices bacon

1. Remove shells from shrimp but leave tails on for "handles".
2. Melt butter in small bowl 1-3/4 minutes in Radarange Oven. Stir chili powder and garlic powder into melted butter.
3. Cut bacon slices in half, lengthwise. Cut each strip into thirds. Dip each shrimp into butter mixture, then roll piece of bacon around shrimp. Place on paper plate lined with paper toweling.
4. Cook in Radarange Oven until bacon is lightly browned, allowing 3-1/2 to 4 minutes for 10-12 shrimp. Be careful not to overcook. Serve hot.

   MICRO-TIP: Lemon wedges and parsley may be used for garnish.

Bacon Wrapped Olives

12 pleasing appetizers.

3 slices bacon

12 pimiento stuffed olives

1. Cut bacon slices crosswise into quarters.
2. Wrap pieces of bacon around well-drained stuffed olives and place on plate lined with paper towel.
3. Cook in Radarange Oven 6-8 minutes.

   MICRO-TIP: Exact time will depend on size of these hors d'oeuvres, but bacon should be crisp and lightly browned. Serve hot.

37
Appetizers and Beverages

Austrian Stuffed Mushrooms

A tray of mouth-watering mushrooms.

1 lb. large fresh mushrooms of uniform size

1. Remove stems from mushrooms. Chop stems.

1/4 lb. finely ground beef
1 minced clove garlic
1/4 cup minced onion
1 tablespoon finely chopped parsley
1/2 teaspoon salt
1 teaspoon Kitchen Bouquet

2. Combine ground beef, chopped mushroom stems, garlic, onion, parsley, salt and Kitchen Bouquet.
3. Cook in Radarange Oven 3 minutes. Stir every 45 seconds.

2 tablespoons lemon juice

4. Dip each mushroom cap in lemon juice and fill with meat mixture.
5. Arrange about 1/4 of mushrooms at one time on plate. Bake in Radarange Oven 4 to 4-1/2 minutes. Serve immediately.

Teriyaki Strips

3 dozen oriental appetizers.

1 lb. 1/4-inch thick, sirloin steak

1. Cut beef into 36 thin strips. Spread in 2-quart utility dish.

1/2 teaspoon ginger
1/2 cup bottled teriyaki sauce
1/4 cup dry sherry

2. Combine ginger, sauce and sherry. Pour over meat strips. Refrigerate 1 hour, basting several times. Drain.
3. Thread meat strips onto bamboo skewers. Arrange in shallow baking dish. Cook in Radarange Oven 2 minutes or until steak is done to taste. Remove meat from skewers to serve.

MICRO-TIP: Serve hot with buttered slices of party rye bread.

Mulled Wine Punch

Sweet-Sour Meat Balls

20 to 24 savory, sweet-sour meat balls.

1 finely chopped onion
3/4 teaspoon seasoned salt
1 (10-3/4 oz.) can tomato soup
3 tablespoons lemon juice
1/4 cup firmly packed brown sugar

1. Combine onion, salt, soup, lemon juice and brown sugar in 9 x 9 x 2-inch glass dish. Cook in Radarange Oven 7 minutes. Stir every 3 minutes.

1 lb. ground beef

2. Roll ground beef into small meat balls. Place meat balls in above sauce. Spoon some sauce over meat balls. Cook in Radarange Oven 7 minutes. Turn halfway through cooking time.

1 (13-1/4 oz.) can drained pineapple chunks

3. Stir in pineapple. Heat 1 additional minute. Serve with toothpicks.

MICRO-TIP: These meat balls are also good served as main dish. Accompany with cooked rice or whipped potatoes, as desired.

Party Tuna Balls

Perfect party pleasers.

1 (6 oz.) can tuna
1 egg
1 cup bread crumbs
2 tablespoons snipped parsley
1/4 cup minced onion
1/2 (10-1/2 oz.) can condensed consommé
1/4 cup mayonnaise
1 tablespoon prepared mustard
1 teaspoon poultry seasoning

1. Flake the tuna. Beat the egg. Combine all ingredients except corn flake crumbs.
2. Chill.

1 cup corn flake crumbs

3. Form mixture into 1-inch balls. Roll in corn flake crumbs. Arrange 12 balls in circle on sheet of waxed paper. Arrange on paper plate.
4. Heat in Radarange Oven 2-1/2 minutes, or until very hot. Serve immediately.

MICRO-TIP: Tuna balls may be prepared ahead and frozen. In this case, defrost 12 balls in Radarange Oven 4 to 4-1/2 minutes.

Hot Cheese Dip

Delicious, yet different, dip.

1/4 large green pepper
1/2 bunch green onions
2 (5 oz. each) jars very sharp cheese spread
1 (7 oz.) can minced, drained, clams
2 to 4 dashes Tabasco
Garlic powder to taste

1. Finely chop green pepper and mince onions. Combine all ingredients in 1-1/2-quart glass casserole.
2. Heat, uncovered, in Radarange Oven 3 minutes, or until cheese melts. Stir after every minute.

MICRO-TIP: May be served very hot with crisp corn chips for dippers.

Oysters Casino

2 dozen seashore shell specials.

2 dozen freshly opened oysters

3 slices bacon
1/4 cup seasoned crumbs
2 tablespoons minced onion
2 tablespoons minced green pepper
2 tablespoons minced parsley
2 tablespoons minced celery
1 tablespoon butter
1 teaspoon Worcestershire sauce
Dash Tabasco

Paprika

1. Place 2 oysters in deep half of one shell. Repeat to fill 12 shells. Arrange filled shells on plastic tray or paper plate.
2. Cook and crumble bacon. Combine with remaining ingredients. Spoon over oysters. Cook in Radarange Oven 4 minutes.
3. Sprinkle with paprika. Serve hot.

MICRO-TIP: Custard cups may be substituted for oyster shells. When using custard cups, place 6 at one time in circular arrangement in Radarange Oven. Remember to reduce timing slightly to accommodate smaller oven load.

Ham Nibbles

20 bite-size nibbles.

1 cup ground or minced cooked ham
3 tablespoons mayonnaise
2 teaspoons prepared mustard

1. Mix together ham, mayonnaise and mustard.

1 pkg. Melba toast rounds
1 can rolled anchovies

2. Spread mixture on toast rounds. Place rolled anchovy in center of each round.
3. Heat 10 "Nibbles" in Radarange Oven 1-1/4 minutes. Serve hot.

Nachos

4 dozen nifty nachos.

1 (1 lb.) can refried beans
1 (5-1/2 oz.) pkg. tortilla chips
4 oz. shredded Cheddar cheese
10 sliced green stuffed olives

1. Spread 1 teaspoon refried beans on each tortilla chip. Top with one teaspoon cheese and thin slice of olive.
2. Place 8 on paper plate. Cook in Radarange Oven 30 seconds. Serve warm.

Rye Savories

4 to 5 dozen rye savories.

8 slices bacon
1/2 lb. sharp cheese
1 medium onion

1. Partially cook bacon in Radarange Oven. Put cheese, onion, and bacon through fine blade of food grinder.

2 teaspoons dry mustard
2 tablespoons mayonnaise
Party rye slices

2. Blend mustard and mayonnaise into cheese mixture. Spread thinly on slices of party rye.
3. Arrange slices on paper plates. Heat in Radarange Oven 35-40 seconds.

Bread Spreads

16 party-perfect treats.

4 slices white bread
2 tablespoons mayonnaise

1. Spread mayonnaise on bread slices.

4 slices cooked turkey or ham
4 slices Cheddar cheese

2. Place meat on bread. Top with cheese. Quarter bread slices. Cook in Radarange Oven 2-1/2 to 3 minutes, or until cheese melts.

MICRO-TIP: Tuna may be substituted for meat.

Appetizers and Beverages

Tangy Tomato Twister

One quart of tangy tomato taste.

4 cups tomato juice
1 cup water
1/2 cup packed celery leaves
1/4 cup diced onion
3 whole cloves
1 bay leaf
1/2 teaspoon salt
1/4 teaspoon pepper

1. Combine ingredients in 2-quart glass dish. Heat in Radarange Oven 15 minutes. Strain before serving.

MICRO-TIP: 1 teaspoon horseradish will give zip to the cocktail.

Hot Eggnog

1-1/2 quarts of Humpty Dumpty's Delight.

1 quart milk

4 egg yolks
1/3 cup sugar
1/4 teaspoon cinnamon
1/4 teaspoon nutmeg
1/4 teaspoon vanilla

1. Heat milk in 2-quart covered glass dish in the Radarange Oven for 5 minutes.
2. Beat egg yolks lightly with sugar, spices and vanilla. Stir one cup of hot milk into egg mixture, then gradually blend all of egg mixture into milk. Return to Radarange Oven for 5 minutes. Stir halfway through cooking time.

Cocoa

3-4 "cocoluscious" servings.

1/4 cup warm water
3 tablespoons sugar
2-1/2 tablespoons cocoa

3 cups milk
Dash salt

1. Place cocoa, sugar and water in 1-1/2-quart glass casserole. Heat in Radarange Oven 30 seconds.
2. Stir in milk and salt. Heat in Radarange Oven 5-5-1/2 minutes.

Mexican Hot Chocolate

A "South of the Border" treat for 4.

1 quart milk
2 oz. unsweetened chocolate
1/4 cup sugar
1/2 teaspoon vanilla
1 teaspoon ground cinnamon
Pinch of salt

1. Combine ingredients in 1-quart measure. Heat in Radarange Oven 5-7 minutes. Stir halfway through cooking time.
2. Remove from Radarange Oven. Beat until well blended and foamy.

Hot Mulled Cider

A fireside treat for 2 or 3.

2 cups apple cider
3 cloves
1-inch piece of stick cinnamon

1. Combine ingredients in 1-quart glass measure. Heat in Radarange Oven 3-1/2-4 minutes. Stir halfway through cooking time.

Hot Chocolate Ukranian Style

5 cupfuls of warmth for the coldest of winter days.

1 (1 oz.) square unsweetened chocolate

1/4 cup sugar
1/8 teaspoon salt
1-1/4 cups boiling water

3/4 cup milk
3/4 cup cream

2 cups freshly brewed coffee
1 teaspoon vanilla

1. Melt chocolate in 1-1/2-quart casserole in Radarange Oven 1 minute
2. Stir in sugar, salt and water. Cook in Radarange Oven 2 minutes.
3. Add milk and cream. Stir, then cook in Radarange Oven 2 minutes.
4. Mix in vanilla and coffee. Cook in Radarange Oven 2 minutes.

Hot Grape Punch

4 purple pleasures.

2 (6 oz. each) cans frozen grape juice
6 (6 oz. each) cans of water

2 tablespoons brown sugar
2 sticks cinnamon
1 teaspoon cloves

1. Dilute grape juice according to directions on can.
2. Stir in remaining ingredients. Cook in Radarange Oven 5-6 minutes. Serve warm.

Irish Coffee

1 cup of Irish cheer.

1 cup strong hot coffee
1 jigger Irish whiskey
1 teaspoon sugar
1 spoonful of whipped cream

Place ingredients in large cup or mug. Stir. Heat in Radarange Oven 1-1/2 to 2 minutes.

Moroccan Mint Tea

4 cups of far-away flavor.

4 cups water

Large bunch fresh mint leaves
1 tablespoon sugar per cup
1/2 teaspoon green tea per cup

1. Boil 4 cups of water in 1-quart glass measure in Radarange Oven.
2. Place mint leaves, sugar and green tea in boiling water. Strain.
3. Pour into individual servings.

Russian Tea

Tea from across the sea.

1 cup powdered orange flavor drink
1 cup sugar
1/2 cup instant tea
1 teaspoon ground cinnamon
1 teaspoon ground cloves

Mix ingredients well. Store in airtight container. Use 2-3 teaspoons per cup of water. Simmer in Radarange Oven 1-1/2 to 2 minutes to blend flavors.

Mulled Wine Punch

17 servings of punch for a party.

2 tablespoons whole cloves
2 tablespoons whole allspice
2 tablespoons broken stick cinnamon
4 cups boiling water

1. Tie spices loosely in piece of cheese cloth. Place in deep casserole with boiling water. Cook in Radarange Oven 5 minutes.

3-4 tablespoons instant tea
1 (6 oz.) can frozen tangerine or orange juice concentrate
1 (6 oz.) can frozen Hawaiian Punch concentrate

2. Remove spices. Add tea and juice concentrates. Cook in Radarange Oven 5 minutes. Stir every minute.

2 (25 oz. each) bottles rosé wine

3. Stir in wine. Serve immediately.

MICRO-TIP: Garnish with lemon or orange slices. Liquid refreshments can always be reheated quickly in Radarange Oven. Allow 1 to 1-1/2 minutes per cup.

Tea - Based Fruit Punch

Time for tea-based fruit punch: 2-1/2 quarts.

1-1/2 cups water
3 tea bags

1. Bring water to boil in Radarange Oven 4 minutes. Immerse tea bags in water. Brew 5 minutes.

1-1/2 cups sugar
1 cup water

2. Dissolve sugar in water in large bowl. Heat in Radarange Oven 2 minutes. Add 6 to 8 ice cubes to cool.

2 cups orange juice
1-1/2 cups lemon juice
1/8 teaspoon salt

3. Mix in tea, fruit juices and salt. Stir well.

1 quart ginger ale

4. Pour ginger ale into punch just before serving.

MICRO-TIP: Place sprig of mint in each glass as it is served.

Hot Rum Lemonade

Come taste this rum!

3/4 cup water
Juice of half lemon
1 jigger rum
1 teaspoon honey

1. Combine ingredients in one cup measure or directly in mug or tea cup.

2. Heat in Radarange Oven 1 to 1-1/2 minutes.

MICRO-TIP: Remember to increase each ingredient proportionately when increasing the number of servings.

45
Fish and Seafood

Whether you're fishing for compliments or just preparing fish, a few guidelines will help you catch compliments. This chapter includes a variety of fish and shellfish which can be prepared in the Radarange Oven. Cooking fish in the Radarange Oven eliminates using huge quantities of water for shellfish, and a cool type of cooking. You'll find recipes for whole, stuffed, fillets, casseroles, lobster, crab, scallops and oysters.

Serve the fillets with one of the sauces found in the sauce section—lemon butter or Bearnaise. Teriyaki or soy sauce and French salad dressing can be used to enhance the flavor and color of fish.

Fish Tips

1. Allow 4 minutes per pound.
2. Cook fish covered to hold in the steam and lessen the cooking time. Plastic wrap or wax paper may be used.
3. Fish should be cooked only until it may be easily flaked with a fork.
4. Cook fish just before planning to serve. It does not reheat as well as other foods and there is a chance of overcooking when reheating.
5. Fish may be cooked without the addition of fats.
6. Allow to stand covered 1-2 minutes. Overcooking will cause dry texture and strong flavor.
7. If browning is desired, use Amana Browning Skillet.

Stuffed Red Snapper

A snappy dish for 6.

1 whole red snapper
Salt
Pepper

1. Rinse prepared whole fish and pat dry. Sprinkle cavity with salt and pepper.

1 egg
1/4 cup evaporated milk
2 tablespoons melted butter or margarine
1-1/2 cups soft bread crumbs
1/2 cup minced celery leaves
1/4 cup minced onion
1/4 cup minced parsley
1 teaspoon salt
1/8 teaspoon seasoned pepper
Pinch dill weed

2. Mix egg, milk, butter, bread crumbs, celery leaves, onion, parsley, salt, pepper and dill weed together to make stuffing.

3. Stuff fish. Skewer with wooden picks. Lace with string to close. Make 3 small slashes in top of fish skin. Place fish on greased heavy brown paper or parchment paper. Fold over. If necessary secure ends with rubber bands.

4. Line bottom of oven with paper towels. Place stuffed fish diagonally in Radarange Oven. Cook 12 minutes. Turn halfway through cooking time. Let stand 5 minutes before serving.

MICRO-TIP: May be garnished with lemon wedges and water cress.

Stuffed Red Snapper

Baked Stuffed Whole Fish

4 servings of your very favorite fish.

1-1/2 lbs. whole fish
1/4 cup melted butter or margarine
1 teaspoon salt

1. Wash dressed whole fish and wipe dry with paper towel. Brush interior of fish with half of butter. Sprinkle with salt.

2 cups bread cubes
1/4 cup boiling water
1/4 cup finely chopped celery
1/4 teaspoon poultry seasoning
2 tablespoons minced onion

2. Toss remaining ingredients together lightly for stuffing. Loosely fill cavity. Fasten with toothpicks.

Paprika

3. Place stuffed fish in suitable size baking dish. Sprinkle with paprika. Cook in Radarange Oven 6-1/2 to 7-1/2 minutes or until fish can be flaked with fork.

MICRO-TIP: Several different types of fish can be used in this recipe: lake trout, white fish and red snapper are just a few of many.

Snappy Stuffed Trout

4 snappy trout.

4 brook trout
1 lemon

1. Moisten interior of trout with juice of lemon.

2 tablespoons melted butter
2 minced shallots
4 slices cooked and crumbled bacon
1-1/2 cups bread crumbs
1/4 cup chopped white raisins
2 tablespoons tomato sauce
1 tablespoon chopped parsley
1/2 teaspoon salt
1/2 teaspoon pepper
1 teaspoon Madeira wine
Dash Tabasco

2. Sauté shallots in butter in 10-inch ceramic skillet in Radarange Oven 3 minutes. Mix in remaining ingredients. Loosely stuff trout. Fasten with toothpicks.

Paprika

3. Sprinkle fish lightly with paprika. Place fish in suitable size glass baking dish. Cook in Radarange Oven 8 minutes or until fish can be flaked with a fork. Turn dish every 2 minutes.

Fish and Seafood

Crusty Salmon Steaks

4 savory salmon steaks.

2 eggs
2 tablespoons half-and-half
1 tablespoon lemon juice
1/2 teaspoon salt

1. In shallow bowl combine eggs, half-and-half, lemon juice and salt. Beat well.

4 (3/4-inch thick) salmon steaks
1 cup packaged cracker meal

2. Dip each steak into egg mixture, then into crumbs.

2 tablespoons butter or margarine

3. Melt butter in 2-quart utility dish in Radarange Oven 1-1/2 minutes. Arrange steaks in dish. Bake in Radarange Oven 10 minutes. Turn dish halfway through cooking time.

Salmon Quiche

A special salmon spectacular for 6.

2 cups cooked, boned salmon chunks
1/2 cup sliced ripe olives
2 tablespoons chopped parsley
9-inch baked pastry shell
1 cup shredded cheddar cheese

1. Lightly mix salmon, olives and parsley. Place in pastry shell. Top with cheese.

3 eggs
1 cup half-and-half
1/4 teaspoon salt
1/4 teaspoon onion powder

2. Combine eggs with half-and-half, salt and onion powder. Pour over salmon. Bake in Radarange Oven 16 minutes. Turn quarter-turn every 4 minutes. Let stand 5 minutes before serving.

Salmon Potato Souffle

Souffle of the day to serve 6.

2 tablespoons butter
1 (16 oz.) can drained salmon
2 tablespoons chopped celery
2 tablespoons chopped parsley
1 tablespoon finely chopped onion
1 tablespoon lemon juice
1 teaspoon Worcestershire sauce
1/4 teaspoon salt

1. Place butter in 10-inch ceramic skillet. Melt in Radarange Oven 15 to 30 seconds. Add salmon, celery, parsley, onion, lemon juice, Worcestershire sauce and salt. Cook, covered, in Radarange Oven 5 minutes. Turn halfway through cooking time.

2 cups seasoned mashed potatoes
3 beaten egg yolks
3 stiffly beaten egg whites
1/2 teaspoon caraway seed

2. Mix in potatoes, egg yolks, egg whites and caraway seed. Return to Radarange Oven 8 minutes. Turn quarter-turns every 2 minutes.

MICRO-TIP: Serve with cream sauce or creamed peas. Tuna may be substituted for salmon.

Fish and Seafood

Creole Halibut

An attractive fish dish for four.

1-1/2 lbs. halibut steaks

1. Arrange halibut in greased 2-quart utility dish. Set aside.

1 cup chopped onion
1/2 cup chopped green pepper
1/4 cup finely chopped celery
1 minced clove garlic
1/4 cup butter or margarine

2. Place onion, green pepper, celery, garlic and butter in small bowl. Cook in Radarange Oven 3 minutes, stir every minute.

1 (1 lb.) can stewed tomatoes
1 (3 oz.) can drained, sliced, broiled mushrooms
1 teaspoon salt
1/2 teaspoon sugar
Dash Tabasco

3. Blend in tomatoes, mushrooms and seasonings. Pour over fish. Cook in Radarange Oven 8 to 10 minutes or until fish flakes easily with fork. Let stand covered 5 minutes before serving.

MICRO-TIP: May be garnished with parsley and lemon wedges.

Halibut Teriyaki

A Japanese fish "fry": 4-6 servings.

1/4 cup soy sauce
1/4 cup dry white wine
1-1/2 tablespoons salad oil
1 tablespoon thinly sliced green onion
2 teaspoons sugar
1/8 teaspoon garlic powder

1. Blend all ingredients except halibut in 1-cup covered jar. Shake vigorously. Repeat shaking 15 minutes later.

1-1/2 lbs. halibut fillets

2. Arrange fish in single layer in 2-quart utility dish. Pour liquid over fish. Cover and refrigerate 30 minutes. Drain all but 2 tablespoons liquid from dish. Cook, covered, in Radarange Oven 4 minutes or until fish flakes easily with a fork.

Fillet of Sole with Oyster Sauce

4 servings of "saucy" sole.

1 (16 oz.) pkg. thawed fillet of sole
1/2 teaspoon salt

1. Separate and arrange fillets in 1-quart glass utility dish. Sprinkle with salt.

1 (10-1/2 oz.) can oyster stew
1 (2 oz.) can sliced, drained mushrooms
Paprika

2. Blend oyster stew with mushrooms. Pour over fillets. Sprinkle with paprika. Cook in Radarange Oven 6 minutes, turn dish halfway through cooking time.

MICRO-TIP: For taste of tartness, add sprinkling of lemon juice.

Creole Halibut

Lemon Sauce for Fish

Tart but tasty 1-1/2 cups.

2 tablespoons butter or margarine
2 tablespoons all-purpose flour
1/2 teaspoon salt

1. Melt butter in 1-quart casserole in Radarange Oven 30 to 45 seconds. Stir in flour and salt.

1 egg yolk
1/2 cup light cream

2. Beat egg yolk into cream. Pour into flour mixture, stirring continually. Cook in Radarange Oven 45 seconds.

1 cup fish broth
1/4 cup lemon juice
2 tablespoons chopped parsley
Dash Tabasco

3. Mix in remaining ingredients. Cook in Radarange Oven 1-1/2 to 2 minutes, stirring every 30 seconds.

Rondeau Sole

4 "soulful" servings.

1 (16 oz.) pkg. thawed sole fillets
Salt
Paprika
Butter

1. Separate and arrange fillets on large plate. Sprinkle with salt and paprika. Generously butter four (5 oz. each) glass custard cups.

1-1/2 cups bread crumbs
1/4 cup minced celery
2 tablespoons melted butter
2 tablespoons hot water
1 tablespoon minced onion
2 teaspoons chopped parsley
1/4 teaspoon seafood seasoning
1/4 teaspoon salt

2. Combine remaining ingredients for filling. Spread over fish. Roll up fillets with stuffing pinwheel style. Set on sides in custard cups. Cook in Radarange Oven 6-1/2 minutes.

French Fish Fillets

Fillets from France for 4.

1 lb. fish fillets
1/4 cup French dressing
1/2 cup cracker crumbs
Paprika

1. Dip fillets in dressing. Coat with crumbs. Place fillets in greased baking dish. Sprinkle with paprika.

2. Bake in Radarange Oven 6 minutes. Turn halfway through cooking time.

Scalloped Corn and Oysters

An original oyster dish to serve 8.

25 crumbled crackers
2 tablespoons melted butter

1. Combine 1/4 of cracker crumbs with butter. Set aside.

1 (1 lb.) can drained whole kernel corn
1 (1 lb.) can cream-style corn
1 (10-1/2 oz.) can oyster stew
1 cup washed, drained, fresh oysters
1/4 cup finely chopped onion
1 teaspoon salt
1/4 teaspoon pepper

2. Combine remaining ingredients except paprika, in glass casserole.

Paprika

3. Bake in Radarange Oven 7 minutes. Stir. Sprinkle reserved crumbs over top. Sprinkle with paprika, if desired. Cook for additional 8 minutes in Radarange Oven. Let stand 4 to 5 minutes before serving.

Shrimp Casserole

Splendid shrimp for 6.

1 (10-1/2 oz.) can cream of mushroom soup
2 tablespoons chopped green pepper
2 tablespoons chopped onion
1 tablespoon lemon juice
2 cups cooked rice
1/2 teaspoon Worcestershire sauce
1/2 teaspoon dry mustard
1/4 teaspoon pepper
1/2 lb. cleaned, uncooked shrimp

1. Combine soup, green pepper, onion, lemon juice, rice, Worcestershire, mustard, pepper and shrimp.

1/4 cup bread crumbs
Butter
Paprika

2. Place in 1-1/2-quart casserole. Top with bread crumbs. Dot with butter. Sprinkle with paprika.

3. Cover and bake in Radarange Oven 10 minutes. Turn dish halfway through baking time. Remove lid for last 2 minutes.

Fish and Seafood

Salmon Loaf

4 slices of salmon.

1 (1 lb.) can salmon
1 egg
3/4 cup bread crumbs
1/2 cup chopped celery
1/4 cup evaporated milk
2 teaspoons onion juice
2 teaspoons lemon juice
1/4 teaspoon salt
Paprika

Flake salmon in large bowl. Mix in all other ingredients. Spoon mixture into 1-quart utility dish. Sprinkle generously with paprika. Cook in Radarange Oven 9 to 10 minutes or until loaf is barely firm.

MICRO-TIP: Garnish loaf with parsley sprigs and wedges of lemon to add color, taste and zest.

Lettuce Steamed Salmon With Tarragon Dressing

Stuffed salmon for six.

1/2 small head iceberg lettuce

2 lbs. dressed fresh salmon
Salt
Pepper

1/2 cup mayonnaise
1/2 cup dairy sour cream
1 tablespoon lemon juice
2 teaspoons finely chopped onion
1-1/2 teaspoons tarragon
1/4 teaspoon salt
1/8 teaspoon pepper
1/8 teaspoon paprika

1. Separate lettuce leaves. Arrange in 2-quart utility dish.
2. Sprinkle cavity of salmon with salt and pepper. Place salmon over lettuce. Cover with plastic wrap. Cook, covered, in Radarange Oven 11 minutes. Turn fish over and turn dish halfway through cooking time. Center portion of fish should flake easily. Chill several hours or overnight, discarding lettuce. Stuff with tarragon dressing.

   Blend mayonnaise with sour cream. Stir in lemon juice, onion and spices. Chill thoroughly.

Cod Fillets Baked In Lime Sauce

Cod from the Cape for 4.

1 lb. thawed cod fillets

2 thinly sliced green onions
1/4 cup lime juice
1 tablespoon butter
1/2 teaspoon salt
1/4 teaspoon powdered ginger
1/8 teaspoon white pepper
Dash powdered bay leaf
Butter

1. Arrange fillets in 1-quart utility dish.
2. Combine onions, lime juice, butter and spices. Stir well. Pour over fillets. Dot with butter. Bake in Radarange Oven 4-1/2 minutes. Turn dish halfway through cooking time.

Striped Bass Stuffed With Shrimp

6 tempting portions of pleasure.

Juice of 2 lemons
1 (5 lb.) striped bass

1. Squeeze juice of lemons over bass. Marinate 1 hour.

1 tablespoon butter
1 cup finely chopped onions
1 mashed clove garlic

2. Melt butter in 9-1/2-inch Amana Browning Skillet. Sauté onions and garlic in Radarange Oven 3 minutes.

1 tablespoon all-purpose flour
1/2 cup beef broth

3. Stir in flour. Cook 1 minute. Pour in beef broth. Cook 1 minute. Remove from Radarange Oven. Cool.

1/2 lb. fresh chopped shrimp
1/2 cup chopped fresh mushrooms
3 beaten egg yolks
1 tablespoon finely chopped parsley
1/2 teaspoon horseradish mustard
1/2 teaspoon salt
1/2 teaspoon pepper
Dash cayenne

4. Blend in shrimp, mushrooms, egg yolks, parsley, mustard, salt, pepper and cayenne. Fill bass cavity with stuffing. Seal with toothpicks.

8 thin slices larding pork

5. Place half of pork slices in bottom of 2-quart utility dish. Arrange bass on pork. Top with remaining slices of pork. Cook in Radarange Oven 16 to 18 minutes. Turn dish every 4 minutes, basting with natural juices. Fish should be fork tender.

1/4 cup melted sweet butter

6. Pour hot melted butter over bass.

Shrimp Sauce

2 cups shrimp sauce for your refreshing summer salad.

1 cup finely diced cooked shrimp
1 cup medium white sauce
1/2 cup dairy sour cream
1 tablespoon chopped capers
1 tablespoon prepared mustard
1 teaspoon curry powder
1/2 teaspoon sugar

1. Mix all ingredients thoroughly in small glass bowl. Prepare white sauce according to directions on page 159.
2. Cover with plastic wrap. Cook in Radarange Oven 1-1/2 minutes. Stir every 30 seconds.

MICRO-TIP: May be chilled and served cold.

Crabmeat Delight

A crabmeat specialty for 4.

1 finely chopped green pepper
1 tablespoon butter or margarine

1. Sauté green pepper in margarine, using 1-1/2-quart casserole. Place in Radarange Oven 3 minutes. Stir halfway through cooking time.

2 cups shredded sharp Cheddar cheese
3/4 cup tomato juice
2 tablespoons all-purpose flour
1/4 teaspoon dry mustard
1/4 teaspoon salt
Dash pepper

2. Mix in 1-1/3 cups cheese, tomato juice, flour, mustard, salt and pepper.

1 slightly beaten egg
3/4 cup scalded milk
2 (7-1/2 oz. each) cans crabmeat

3. Stir in egg. Cook in Radarange Oven 2 minutes. Pour in milk. Mix in crabmeat. Sprinkle remaining cheese on top. Bake 4 minutes. Turn dish halfway through cooking time. Let stand 5 minutes before serving.

Texas Crab Casserole

4 Texas-size servings.

1 cup grated sharp Cheddar cheese
2 cups medium white sauce

1. Stir cheese into warmed white sauce using 1-1/2-quart casserole. Cook in Radarange Oven 30 seconds or until cheese is completely melted. Stir every 10 seconds. Reserve 1 cup.

2 beaten egg yolks
2 cups crabmeat
1/4 cup toasted crumbs
2 tablespoons grated Parmesan cheese
2 tablespoons butter or margarine

2. Blend egg yolks into cheese sauce until smooth. Stir in crabmeat. Sprinkle with crumbs and cheese. Dot with Butter. Top with reserve cheese sauce.

3. Cook in Radarange Oven 8 to 9 minutes. Turn dish every 2 minutes.

Curried Scallops

3 to 4 scoops of scallops.

1 tablespoon butter or margarine
1 lb. rinsed and drained scallops
1/4 cup sliced green onions

1. Sauté onions and scallops in butter in 9-inch pie plate for 4 minutes in Radarange Oven. Sitr every 1-1/2 minutes. Reserve liquid.

2 tablespoons accumulated liquid
1-1/2 teaspoons cornstarch
1-1/2 teaspoons curry powder
1/4 teaspoon salt

2. Combine cornstarch, curry powder and salt with reserve liquid. Pour over fish. Cook in Radarange Oven 2 minutes until sauce is thickened and smooth.

MICRO-TIP: Serve scallops over rice, garnish with tomatoes and watercress.

Clams in the Shell

You'll "dig" these clams.

8 clams

1. Wash and scrub clams with brush in cold water.
2. Place 4 clams in circle on glass pie plate. Cover pie plate with plastic wrap.
3. Place in Radarange Oven 2-1/2 to 3 minutes, or until clams have relaxed and opened.
4. Repeat above procedure with remaining clams.

MICRO-TIP: Delicious served with butter sauce.

Oyster Pepper Sauté

4 sautéed servings.

2 tablespoons butter or margarine
1 tablespoon finely chopped onion
1 chopped green pepper

1/4 lb. sliced mushrooms

18 small drained oysters
1 finely chopped canned pimiento
2 tablespoons dry red wine (optional)

1. Melt 1/2 tablespoon of butter in 9-inch round glass dish in Radarange Oven 30 seconds. Stir in onion and green pepper. Cook 2 minutes. Remove from glass dish and set aside.
2. Using same glass dish, melt 1/2 tablespoon more of butter for 30 seconds in Radarange Oven. Stir in mushrooms and cook for additional 30 seconds. Remove mushrooms and set aside.
3. Melt remaining tablespoon of butter for 45 seconds in same glass dish in Radarange Oven. Place oysters in single layer in dish. Cover with paper towel and cook in Radarange Oven 1-1/2 minutes. Stir in pepper and onion mixture, mushrooms, pimiento and wine (if used). Cook 1-1/2 minutes in Radarange Oven.
4. Using slotted spoon, remove oysters and vegetables to warm serving dish. Replace dish with juices in Radarange Oven and cook about 4 minutes, or until liquid is reduced by almost half. Pour juices over oysters.

MICRO-TIP: May be served over toast or rice.

Salmon Stuffed Green Peppers

4 peppers to beautify your table.

4 large green peppers

1. Cut slice from upper third of each pepper to make scalloped edge. Finely dice slices. Remove seeds and membranes. Parboil peppers in Radarange Oven 5 minutes. Drain.

1 (1 lb.) can drained and flaked salmon
1 cup soft bread crumbs
1/2 cup finely diced celery
1/3 cup mayonnaise
1 egg
2 tablespoons lemon juice
2 tablespoons prepared mustard
2 tablespoons soft butter
1 tablespoon minced onion
1/4 teaspoon salt
1/8 teaspoon tabasco

2. Thoroughly mix diced green pepper with remaining ingredients, except cheese. Fill pepper shells with salmon mixture.

3. Place stuffed peppers upright in greased shallow baking dish. Cook in Radarange Oven 10 to 12 minutes, turn dish every 4 minutes.

2 slices process cheese cut in strips

4. Arrange cheese strips on each pepper.

Salmon Green Bean Casserole

4 servings of a green bean Great!

1 (9 oz.) pkg. frozen Italian green beans
1 (8 oz.) pkg. frozen artichoke hearts

1. Cook beans and artichoke hearts separately in Radarange Oven as directed in Vegetable section. Drain well and reserve.

2 cups cooked, boned salmon chunks

2. Lightly mix vegetables with salmon chunks. Spread salmon mixture in 1-1/2-quart casserole.

1/2 cup Hollandaise sauce
1/3 cup dairy sour cream
1/2 teaspoon shredded lemon peel
1/4 teaspoon crushed tarragon
2 tablespoons toasted, slivered almonds

3. Blend Hollandaise sauce, sour cream, lemon peel and tarragon. Spread over salmon. Sprinkle with almonds. Bake in Radarange Oven 7 minutes until sauce is bubbly.

Salmon Stuffed Green Pepper

Shrimp Creole

4 to 5 servings of a colorful creole dish.

2 tablespoons butter or margarine

1. Melt butter in Radarange Oven in 2-1/2-quart casserole about 1-1/2 minutes.

3/4 cup chopped green pepper
1 cup chopped onion
1 cup chopped celery

2. Stir in green pepper, onion and celery. Cook 4 minutes in Radarange Oven, or until vegetables are barely tender. Stir halfway through cooking time.

1-1/2 tablespoons all-purpose flour
1 (14-1/2 oz.) can tomatoes

3. Sprinkle vegetables with flour and stir to blend. Mix in tomatoes and cook 3 minutes in Radarange Oven. Stir and cook additional 3 minutes.

1 teaspoon sugar
5-6 drops Tabasco
Dash powdered bay leaf
1 teaspoon salt
1/8 teaspoon pepper
1 lb. shelled and cleaned shrimp

4. Blend in sugar, Tabasco, bay leaf, salt and pepper. Stir in shrimp. Heat about 6 minutes in Radarange Oven. Stir every 2 minutes.

MICRO-TIP: May be served over cooked rice.

Curried Shrimp and Broccoli

6 to 8 servings of a shrimp special.

1 tablespoon salt
3 quarts boiling water
4 cups fine egg noodles

1. Pour salt into boiling water. Stir in noodles and cook until tender. Drain.

1 (10-1/2 oz.) can cream of celery soup
1 to 2 teaspoons curry powder
2 teaspoons Worcestershire sauce
1/2 cup finely chopped onion
1/2 cup skim milk

2. Combine soup, curry, Worcestershire and onion in dish. Stir in milk.

3. Cook in Radarange Oven 5 minutes until onion is tender. Stir every 2 minutes.

2 lbs. cooked and cleaned shrimp
1 (10 oz.) pkg. frozen, chopped, cooked broccoli

4. Stir in shrimp and broccoli. Heat 5 minutes, stirring every 2 minutes.

Salt
1 tablespoon shredded Parmesan cheese

5. Gently combine soup mixture with noodles. Salt to taste and toss. Sprinkle with cheese before serving.

Oriental Shrimp

4 servings of spicy shrimp.

12 oz. frozen, raw, peeled, deveined shrimp
3/4 teaspoon chicken stock base
1/2 cup boiling water

1. Defrost frozen shrimp in Radarange Oven. Dissolve chicken stock base in boiling water.

4 tablespoons salad oil
1 crushed clove garlic
1/2 teaspoon grated ginger root

2. Preheat 2 tablespoons oil, the garlic and ginger in 9-1/2-inch Amana Browning Skillet for 4 minutes in Radarange Oven. Remove garlic and ginger. Stir in shrimp. Toss to coat with oil. Cook 2 minutes in Radarange Oven. Stir and remove from skillet to dish.

1 cup celery pieces
1 medium onion

3. Place 2 tablespoons oil in Amana Browning Skillet. Preheat 2-1/2 minutes. Cut onion in eighths. Cut celery in diagonally-shaped pieces. Stir in celery and onion. Cook 4 minutes. Stir halfway through cooking time.

1 green pepper

4. Cut green pepper into chunks. Blend green pepper and shrimp with celery-onion mixture.

1 (6 oz.) can tomato sauce
1 tablespoon sugar
1/4 teaspoon salt
1-1/2 tablespoons cornstarch
1/2 teaspoon MSG
1/4 cup soy sauce

5. Mix together tomato sauce, sugar, salt, cornstarch, MSG, soy sauce and chicken stock. Stir into shrimp mixture.

2 tomatoes

6. Cook, covered, in Radarange Oven 5 minutes. Stir halfway through cooking time. Cut tomatoes in eighths and blend in. Cook 1 additional minute.

MICRO-TIP: May be served with hot rice. Also, 1/4 teaspoon dry ground ginger may be substituted for the ginger root.

Creamed Tuna on Chinese Noodles

Tasty tuna treat for 4.

2 tablespoons butter
1/2 cup finely chopped celery

2 tablespoons all-purpose flour
1/4 teaspoon salt
1/8 teaspoon paprika
1/2 teaspoon Worcestershire sauce
1 cup milk

1 (6-1/2 oz.) can tuna
1 teaspoon lemon juice
1 (3 oz.) can Chinese noodles

1. Place butter and celery in 1-quart glass casserole. Cook in Radarang Oven 2 minutes.
2. Blend in flour, salt, paprika and Worcestershire. Gradually stir in milk. Cook, uncovered, 4 minutes. Stir every 30 seconds.
3. Drain liquid from tuna and flake tuna with fork. Stir tuna into sauce. Stir in lemon juice and heat in Radarange Oven 1-1/2 minutes longer. Serve over crisp Chinese noodles.

Tuna-Rice Casserole

Tuna and rice combination to serve 8.

1/4 cup chopped onion
1 chopped green pepper
6 tablespoons butter

1/4 cup all-purpose flour
1 cup milk

2 cups cooked rice

2 cans white tuna or
1 can tuna and 1 large can boned chicken
1 (10-1/2 oz.) can cream of chicken soup
1 (10-1/2 oz.) can cream of mushroom soup
1 (2 oz.) jar pimientos
1 (4 oz.) can drained mushrooms
Salt (optional)

1. Sauté onion and green pepper in butter using 10-inch ceramic skillet for 5 minutes in Radarange Oven.
2. Blend in flour. Stir in milk gradually until smooth. Return to Radarange Oven for 2 minutes. Stir every 45 seconds.
3. Mix rice into sauce and stir until completely coated.
4. Stir in tuna, soups, pimientos, mushrooms and salt if desired. Cook in Radarange Oven 10 minutes. Turn every 2-1/2 minutes.

Tuna Casserole

Truly tremendous tuna for four.

1 (4 oz.) can shoestring potatoes
1 (10-1/2 oz.) can cream of mushroom soup
1 (7 oz.) can drained tuna
1 (6 oz.) can evaporated milk
1 (3 oz.) can broiled, sliced, drained mushrooms
1/4 cup chopped pimiento

1. Reserve 1 cup of shoestring potatoes for topping. Combine remaining potatoes with other ingredients in 1-1/2-quart casserole. Sprinkle reserved shoestring potatoes on top.
2. Bake in Radarange Oven 10 minutes. Turn dish every 3 minutes.

Tuna Tetrazzini

6 servings of tempting tuna.

- 4 oz. spaghetti
- 1 (10-1/2 oz.) can cream of mushroom soup
- 1/2 cup milk
- 1 (7 oz.) can drained, flaked tuna
- 1 (3 oz.) can sliced, drained mushrooms
- 1/3 cup chopped onion
- 1 cup shredded Cheddar cheese

1. Break spaghetti into 2-inch lengths. Cook and drain.
2. Stir in soup and milk. Blend in tuna, mushrooms, onion and 1/2 cup cheese. Mix lightly.
3. Cover and place in Radarange Oven. Cook 5 minutes. Stir. Heat another 3 minutes. Sprinkle remaining cheese on top. Return to Radarange Oven, uncovered, for additional 2 minutes.

Fish 'N Chips Casserole

Fix this fish dish for 4.

- 1 (10-3/4 oz.) can condensed cream of celery soup
- 1/3 cup light cream
- 1 (3 oz.) can sliced broiled mushrooms and broth
- 1 (2 oz.) bag crushed potato chips
- 1 (7 oz.) can drained tuna

1. Mix together soup, cream, mushrooms and mushroom broth.
2. Arrange layers of crushed chips, flaked tuna and soup mixture in 1-quart casserole, beginning and ending with crushed chips.
3. Cook in Radarange Oven 8 minutes. Turn dish quarter-turn every 3 minutes.

Fish Casserole Florentine

Fancy fish casserole for 4.

- 1 (15-1/2 oz.) can salmon
- 2 cups cooked, drained spinach
- 2 cups thick white sauce
- 2 cups shredded Cheddar cheese
- 1/2 cup toasted buttered crumbs

1. Drain and flake salmon. Chop spinach. Arrange alternate layers of salmon and spinach in greased 1-quart glass casserole.
2. Prepare white sauce according to directions on page 159. Stir cheese into white sauce.
3. Pour hot sauce into casserole. Lift layers of spinach and salmon with fork to distribute sauce. Top with crumbs.
4. Bake in Radarange Oven 5 to 6 minutes. Turn dish every 2 minutes. Let stand 5 minutes before serving.

Tuna-Noodle Medley

Use your "noodle". Serve this medley; 4 to 6 servings.

1 (7 oz.) can tuna
2 cups cooked noodles

1. Drain tuna and flake with fork. Mix noodles and tuna in 1-1/2-quart casserole.

1 (10-1/2 oz.) can mushroom soup
1 teaspoon Worcestershire sauce
2 to 3 tablespoons buttered cracker crumbs
Paprika

2. Mix in soup and Worcestershire sauce. Top with buttered crumbs. Sprinkle with paprika.
3. Bake in Radarange Oven 8 to 10 minutes. Turn dish halfway through cooking time.

Tuna Vegetable Dish

Delightful dish for dinner: 6 servings.

1 (10-1/2 oz.) can cream of mushroom soup
2 tablespoons all-purpose flour
2 tablespoons instant minced onion
1 (13 oz.) can drained tuna
2 tablespoons lemon juice

1. Combine soup, flour, onion, tuna and lemon juice in 1-1/2-quart casserole.

1 (16 oz.) can peas and carrots

2. Drain canned vegetables, reserving 1/4 cup liquid. Add vegetables and reserved liquid to soup mixture.

Potato chips
Paprika

3. Crumble potato chips on top. Sprinkle with paprika. Bake in Radarange Oven 10 minutes. Turn dish halfway through cooking time.

Hawaiian Style Shrimp Curry

Serve happy Hawaiian shrimp for 4.

6 tablespoons butter or margarine
1/4 cup minced onion

1. Place butter and onion in 2-1/2-quart casserole. Cook in Radarange Oven 2 minutes.

1/4 cup all-purpose flour
1 teaspoon curry powder
1 cup milk
1 cup coconut milk

2. Stir in flour and curry powder to make smooth paste. Add milks. Stir. Cook in Radarange Oven 3-1/2 minutes. Stir every minute.

1 tablespoon grated ginger root or 1/2 teaspoon powdered ginger
2 teaspoons lemon juice
1 lb. fresh shrimp

3. Blend in remaining ingredients. Mix well. Cook in Radarange Oven 4 minutes. Stir halfway through cooking time.

MICRO-TIP: May be served with fluffy rice and flaked coconut.

63
Poultry

In this chapter you'll find recipes for poultry prepared in a variety of different ways in the Radarange Microwave Oven. Cornish Cassoulet, Country Club Turkey and Duckling Bordeaux are just a few of the prize-winning recipes. You'll find that poultry cooked in the Radarange Oven is much juicier than when prepared conventionally.

Herbs such as tarragon, curry, saffron and sage complement poultry. The next time you have leftover chicken, make a chicken salad and look for an appropriate herb in our handy herb chart inside the front and back covers.

Poultry Tips

* Allow 7 minutes per lb. for poultry.
* Pierce the skin before placing in the Radarange Oven to prevent popping.
* Seasoned coating mixes and paprika will improve the color of the bird.
* A 10-15 minute standing time before carving helps the juices to set in the meat, and to keep the meat more moist.
* After poultry has started to brown, cover the wing tips, the narrow part of the legs and the high point of the breast bone with small strips of foil. This small amount of foil will slow the cooking of these areas, but will not damage the unit.
* A non-salted vegetable oil is best for basting poultry.
* The Radarange Oven will hold as much as an 18-22 lb. turkey. You may prefer to cook your turkey in a brown paper bag. We recommend stuffing the cavity with a bread dressing, placing the bird in a large brown paper bag, closing the bag with a rubber band, and placing the bird on a large platter in Radarange Oven. The bird should be turned several times during cooking so that each side is up as well as breast-up and breast-down.
* We recommend basting off the juices as they accumulate because the microwaves will be attracted to the liquid, resulting in a longer cooking time.

Chicken Breasts with Artichoke Hearts

4 chicken breasts sure to reach the heart of your appetite.

1-1/2 lbs. whole boned chicken breasts
1 cup packaged bread coating mix

1. Cut chicken breasts in half. Dip in crumb coating. Preheat 9-1/2-inch Amana Browning Skillet in Radarange Oven 4-1/2 minutes.

1 tablespoon butter or margarine

2. Place butter and chicken pieces in skillet. Cook in Radarange Oven 4 minutes. Turn pieces over.

1 (8 oz.) pkg. frozen artichoke hearts
1/4 cup white wine

3. Place artichoke hearts around chicken breasts. Pour in wine. Cook, covered, 2 minutes. Turn dish halfway through cooking time. Chicken should be fork-tender.

2 strips cooked and crumbled bacon
Salt
Pepper

4. Top with crumbled bacon. Season with salt and pepper to taste.

Scalloped Ham, Chicken & Mushrooms

Brunch for 8.

1/4 cup butter

1. Melt butter in 2-quart casserole in Radarange Oven 1 minute.

1/4 cup all-purpose flour

2. Blend in flour. Cook in Radarange Oven 1 minute.

1 cup chicken broth
1 cup light cream

3. Gradually stir in liquids. Cook in Radarange Oven 5 minutes. Stir every minute.

2 cups cut-up, cooked chicken
2 cups diced ham
1 medium chopped green pepper
1/2 cup fresh sliced mushrooms

4. Fold in remaining ingredients. Cook 4 minutes.

MICRO-TIP: May be served over split and toasted English muffins.

Poulet De La Creme

French treat for 4.

1 cup finely chopped celery
1 tablespoon butter

1. Sauté celery in butter using 1-1/2-quart covered casserole in Radarange Oven for 5 minutes.

1 (6 oz.) pkg. Noodles Almondine
2 cups boiling water

2. Blend in noodles and sauce mix. Pour water over noodles and stir.

1 cup diced, cooked chicken
1 cup shredded process cheese
Almonds

3. Stir in chicken and cheese. Mix thoroughly. Cover, and bake 12 minutes in Radarange Oven. Stir every 5 minutes. Let stand, covered, 10 minutes, before serving. Sprinkle with almonds.

Chicken Breast with Artichokes

Chicken Cacciatore

6 servings of chewy chicken.

1/4 cup all-purpose flour
1 teaspoon salt
1/2 teaspoon pepper
1 cut-up broiler-fryer

1. Combine flour, 1 teaspoon salt and pepper. Coat chicken pieces with this mixture.

1/3 cup vegetable oil

2. Preheat 9-1/2-inch Amana Browning Skillet in Radarange Oven 4-1/2 minutes. Place oil in skillet. Brown chicken in hot oil, turning to brown all sides.

4 sprigs parsley
3 sliced medium onions
2 diced pimientos
1 clove garlic
1/4 teaspoon basil
1/4 teaspoon saffron
1 bay leaf

3. Remove chicken. Combine parsley, onions, pimientos, garlic and herbs in Amana Browning Skillet. Cook in Radarange Oven 5 minutes. Remove garlic.

1 (20 oz.) can drained Italian tomatoes
1 teaspoon salt

4. Return chicken pieces to Amana Browning Skillet. Pour tomatoes over chicken and blend in 1 teaspoon salt.
5. Cover, and cook in Radarange Oven about 24-27 minutes. The exact time will depend upon age and tenderness of chicken.

Apricot Baked Chicken

8 to 12 servings of appetizing apricot baked chicken.

5 lbs. chicken fryer pieces

1. Arrange chicken pieces in two, 2-quart glass utility dishes. Place thickest meaty pieces around edges of dishes.

1/4 cup mayonnaise
1 envelope dry onion soup mix
1/2 cup bottled Russian dressing
1 cup apricot preserves

2. Combine mayonnaise, soup mix, Russian dressing and apricot preserves. Spread over chicken, coating each piece.
3. Cover each dish with waxed paper. Bake each dish of chicken separately in the Radarange Oven. Cook 20 minutes per dish. Turn each halfway through cooking time.

Chicken And Papaya Polynesian

A Polynesian pleasure for 6.

2 tablespoons butter or margarine
6 split and boned whole chicken breasts

1. Melt butter in 2-quart utility dish in Radarange Oven 2 minutes. Tip dish to distribute melted butter evenly. Arrange chicken breasts, skin side up, in butter.

1/3 cup thawed frozen pineapple-orange juice concentrate
1 teaspoon soy sauce
1/2 teaspoon ground ginger
1/8 teaspoon garlic powder
Paprika

2. Mix together juice concentrate, soy sauce, ginger and garlic powder to make sauce. Brush sauce generously over chicken. Reserve remaining sauce. Sprinkle chicken evenly with light coating of paprika.

1 peeled, halved papaya

3. Remove seeds from papaya. Cut papaya lengthwise into 1/2-inch-thick slices.
4. Bake chicken in Radarange Oven for 16 minutes, until tender. Turn dish halfway through cooking time. Arrange sliced papaya over chicken prior to last 3 minutes of cooking time.

1 teaspoon cornstarch

5. Blend reserved sauce with cornstarch, until smooth, in small glass bowl. When chicken is cooked, draw off juices with baster. Stir juices into cornstarch mixture.
6. Cook sauce in Radarange Oven until thickened and clear, or about 2-1/2 minutes. Stir in any juices that accumulated around chicken while sauce was cooking.

3 cups cooked rice

7. Spoon sauce over chicken and papaya. Serve with rice.

MICRO-TIP: Lime wedges may also be served with this dish.

Classic Chicken Divan

Divine chicken divan for 6.

2 (10 oz. each) pkgs. frozen chopped broccoli

1. Cook broccoli. Drain.

1/4 cup butter or margarine
1/4 cup all-purpose flour

2. Melt butter in large glass measure in Radarange Oven 30 seconds. Stir in flour until smooth.

2 cups chicken broth
1/2 cup whipping cream
3 tablespoons cooking sherry
1/2 teaspoon salt
1/4 teaspoon pepper

3. Gradually pour in chicken broth, blending thoroughly. Cook in Radarange Oven 5 minutes. Stir every minute. Blend in cream, sherry, salt and pepper.

1/4 cup grated Parmesan cheese
3 cooked, thinly sliced chicken breasts
Paprika

4. Place drained broccoli in large utility dish. Pour half of sauce over broccoli. Stir cheese, reserving 1 tablespoon, into remaining sauce. Place chicken pieces over broccoli. Cover with cheese sauce. Sprinkle remaining cheese and paprika over top. Bake in Radarange Oven 15 minutes. Turn dish every 5 minutes.

Chicken And Onions In Wine

Wine and dine dish for 6.

1 tablespoon vegetable oil
1 cut-up (3 lb.) broiler-fryer

1. Preheat 9-1/2-inch Amana Browning Skillet in Radarange Oven 4-1/2 minutes. Place oil and chicken pieces in skillet. Brown chicken. Drain excess fat.

2 thinly sliced large onions
1/2 cup consomme or bouillon cube and water
1/4 cup dry sherry
1/4 teaspoon powdered thyme
Salt
Pepper

2. Arrange sliced onions over chicken. Add remaining ingredients. Cook in Radarange Oven 25 minutes. Turn chicken every 5 minutes. Season with salt and pepper before serving.

Cranberry Chicken Exotic

4 exotic servings.

3 lb. cut-up broiler-fryer chicken
Salt
Pepper

1. Lightly season chicken with salt and pepper. Preheat 9-1/2-inch Amana Browning Skillet in Radarange Oven 4-1/2 minutes. Brown chicken in skillet 2 minutes. Turn chicken halfway through cooking time.

1/2 cup whole cranberry sauce
2 tablespoons water
2 tablespoons vinegar
1-1/2 tablespoons butter
3 whole cloves
1-inch piece stick cinnamon

2. Combine cranberry sauce, water, vinegar, butter, cloves and cinnamon stick. Pour over chicken. Cook in Radarange Oven 25 minutes. Turn pieces approximately every 8 minutes.

Chicken Dee-lish

4 delicious servings of what the name implies.

1/3 cup chopped onion
1/3 cup sliced celery
1/4 teaspoon allspice
2 tablespoons butter or margarine

1. Sauté onion, celery and allspice in butter in 10-inch ceramic skillet 3 minutes in Radarange Oven.

1-1/2 cups diced, cooked chicken
1 (10 oz.) can chicken gravy
1/2 cup mandarin oranges

2. Stir in chicken, gravy and oranges. Cook 6 minutes in Radarange Oven. Stir every 2 minutes.

MICRO-TIP: Serve with rice sprinkled with nuts.

Far Eastern Chicken

For far eastern flavor try this 4-serving dish.

2-3 lbs. cut-up fryer chicken

1. Arrange chicken pieces in 2-quart utility dish, placing larger pieces around outer edges.

1 cup chopped onion
1 sliced green pepper
1 large chopped tomato

2. Position vegetables around chicken.

1-1/2 teaspoons curry powder
1-1/2 teaspoons cumin
1 teaspoon salt
1/2 teaspoon pepper
1/4 teaspoon cinnamon
1/4 teaspoon garlic powder

3. Combine spices well. Sprinkle over chicken and vegetables. Chill in refrigerator 1 hour.

1/2 cup water or chicken stock

4. Pour water, or stock, over chicken and vegetables. Cook in Radarange Oven 21 to 23 minutes. Turn chicken over after 10 minutes.

Chicken Teriyaki

2 marinaded portions.

1/2 cup soy sauce
1/4 cup dry white wine
1 crushed clove garlic
2 tablespoons sugar
1/2 teaspoon powdered ginger

1. Combine soy sauce, wine, garlic, sugar and ginger. Stir well.

2-1/2 lb. frying chicken

2. Place chicken in heavy duty plastic bag. Pour marinade over chicken. Totally saturate chicken. Tie securely. Refrigerate 1 to 2 hours. Transfer chicken to baking dish. Cook in Radarange Oven 22 to 25 minutes. Allow 5 minute standing period.

Cornish Cassoulet

A delightful dinner for 4!

2 tablespoons vegetable oil
2 (1-1/2 lbs.) Rock Cornish hens
1/2 teaspoon seasoned pepper

1. Preheat 9-1/2-inch Amana Browning Skillet in Radarange Oven 4-1/2 minutes. Add oil and Cornish hens rubbed with pepper. Cover, and cook in Radarange Oven 5 minutes, turning halfway through cooking time.
2. Remove hens from skillet. Keep warm.

1/2 lb. cut, sweet Italian sausage
3 tart apples
1 diced green pepper
1/2 cup chopped onion

3. Add sausage, apples, green pepper and onion to skillet. Cook in Radarange Oven 5 minutes.

1 (1 lb. 4 oz.) can drained white kidney beans
1 teaspoon Worcestershire sauce
1/4 teaspoon Tabasco

4. Stir in beans and seasonings. Place hens on top of bean mixture. Cook in Radarange Oven 10-12 minutes, turning skillet every 2 minutes.

Cornish Hen

1 hen to highlight your table.

1 Rock Cornish game hen

1. Place Cornish hen on cooking grill.

2 tablespoons pineapple preserves
1 tablespoon catsup
Dash of soy sauce
Dash of garlic salt
Dash of onion salt

2. Make glaze by combining preserves, catsup, sauce and salts. Cook in Radarange Oven for 4 minutes. Turn bird over and baste with remaining glaze. Cook 2 to 4 minutes more. Check internal temperature with meat thermometer in thickest part of thigh. It should register 185° internal temperature.

Chicken Broccoli Au Gratin

Chicken, broccoli and cheese to please 4.

Whole cooked chicken
2 (10 oz. each) pkgs. frozen broccoli

1. Slice chicken and remove skin. Cook broccoli according to directions in vegetable section.

4 tablespoons butter
4 tablespoons all-purpose flour
Salt to taste
1/4 teaspoon white pepper

2. Melt butter in 1-quart glass casserole about 1 minute in Radarange Oven. Blend in flour and seasonings.

2 cups chicken stock
1/2 cup heavy cream

3. Gradually stir in chicken stock until smooth. Cook 4 to 5 minutes in Radarange Oven. Stir every minute until sauce is thickened. Whip cream and fold in.

1/4 cup grated Parmesan cheese

4. Arrange broccoli in glass utility dish. Cover with half of sauce. Lay sliced chicken over broccoli and sauce. Stir cheese into remaining sauce. Pour over chicken. Cook in Radarange Oven 2 minutes.

Curried Chicken In Avocado Halves

A California flavor for 4.

1/3 cup chopped, pared apple
1/4 cup chopped onion
1/4 cup chopped celery
Dash garlic powder
1-1/2 teaspoons curry powder
2 tablespoons all-purpose flour

1. Measure apple, onion, celery, garlic powder, curry powder and flour into 1-quart sauce dish. Mix well.

1 cup chicken broth
1/2 teaspoon salt
Dash pepper

2. Gradually stir in broth. Blend in salt and pepper. Cook in Radarange Oven 4 to 4-1/2 minutes until thickened. Stir every minute.

1-1/2 cups cooked, diced chicken

3. Stir in chicken. Cook 1-1/2 minutes in Radarange Oven.

2 large avocados
2 cups cooked rice

4. Halve and peel avocados. Arrange avocado halves on cooked rice in shallow baking dish. Heat in Radarange Oven 1-1/2 to 2 minutes, or until rice and avocados are hot.
5. Spoon hot curried chicken over avocado halves.

MICRO-TIP: May be served with choice of condiments:
Chopped hard cooked egg
Crumbled cooked bacon
Sweet mixed pickle
Coconut
Raisins
Chutney
Preserved ginger
Chopped peanuts

Hot Chicken Salad

4 servings of super salad.

2 cups diced, cooked chicken or turkey
1 cup thinly sliced celery
1/2 cup chopped, salted peanuts
1/2 teaspoon salt
2 teaspoons grated onion
1 cup mayonnaise
2 tablespoons lemon juice

1. Combine chicken, celery, peanuts, salt, onion, mayonnaise and lemon juice. Spoon lightly into 4 individual 6-ounce glass dishes.

1/2 cup shredded Cheddar cheese
1 cup crushed potato chips

2. Sprinkle with cheese and crushed potato chips. Bake in Radarange Oven 4-1/2 to 5 minutes, or until mixture is well-heated.

Chicken A La King

8 royal servings.

1/3 cup chopped green pepper
1 teaspoon minced onion
2 tablespoons melted butter

1. Using 2-quart casserole, sauté onion and green pepper in butter 2 to 3 minutes in Radarange Oven.

3/4 cup all-purpose flour
3-1/2 cups chicken broth

2. Add flour, and mix well. Gradually blend in broth. Stir until smooth. Cook in Radarange Oven 4 minutes or until thick. Stir every minute.

3 cups diced, cooked chicken
1 cup chopped, cooked mushrooms
1/3 cup chopped pimientos
1-1/2 teaspoons seasoning salt
1/4 teaspoon pepper

3. Stir in chopped chicken, mushrooms, pimientos and seasonings.

2 egg yolks
3/4 cup milk

4. Blend in egg yolks and milk. Cook in Radarange Oven 10 minutes. Stir every 2 minutes.

MICRO-TIP: Delicious when served over biscuits or bread prepared in Radarange Oven.

Turkey Ham Eleganté

6 elegant portions.

1-1/2 tablespoons butter or margarine
1/2 cup bread crumbs

1. In small bowl, melt 1-1/2 tablespoons butter in Radarange Oven 30 seconds. Mix with bread crumbs. Set aside.

2 tablespoons butter or margarine
1/2 cup chopped onion
3 tablespoons all-purpose flour
1/2 teaspoon salt
1/8 teaspoon pepper

2. In 2-quart glass casserole, melt 2 tablespoons butter in Radarange Oven 1 minute. Stir in onion and cook 1 minute more. Blend in flour, salt and pepper.

1 (3 oz.) can undrained, sliced mushrooms
1 cup light cream
2 tablespoons dry sherry
2 cups cubed, cooked turkey
1 cup cubed, cooked ham
1 (5 oz.) can drained and sliced water chestnuts

3. Stir in mushrooms, cream and sherry. Cook in Radarange Oven 1 minute. Stir well. Cook 3 additional minutes. Gently stir in turkey, ham and water chestnuts. Cook in Radarange Oven 7 minutes, stirring every minute.

1/2 cup shredded process Swiss cheese
Paprika

4. Top with cheese and wreath with buttered crumbs. Return to Radarange Oven for 1 minute. Sprinkle with paprika.

Turkey Ham Elegante

After Holiday Casserole

6 servings to add to your holiday menu.

1 cup instant rice
2 tablespoons instant minced onion

1. Prepare rice according to directions on box, adding minced onion to boiling water.

1/2 (5 oz.) pkg. frozen green peas
1/2 cup finely diced green pepper
6 slices turkey

2. Fluff rice with fork and spread in 1-1/2-quart utility dish. Defrost peas and drain. Sprinkle peas and green pepper over rice. Cover with turkey.

1 (10-3/4 oz.) can cheese soup
1 cup milk

3. Mix soup with milk and pour over turkey.

1 cup finely crushed cheese crackers

4. Combine crumbs with butter and sprinkle on top of casserole.

3 tablespoons melted butter

5. Cover dish with plastic film and cook in Radarange Oven 12 minutes, turning dish every 3 minutes. Let stand, covered, 5 minutes before serving.

Chicken Casserole

4-6 servings of a family favorite!

6 slices bread

1. Cube 2 slices of bread. Place in greased 2-quart casserole.

2 cups cooked chicken or turkey
1/2 cup chopped celery
1/4 cup chopped onion
1/4 cup chopped pimiento
1/2 cup mayonnaise
1/2 teaspoon salt
1/8 teaspoon pepper

2. Combine chicken, onion, celery, pimiento, mayonnaise and seasonings. Pour mixture over bread. Remove crust from remaining bread. Place over casserole.

2 eggs
1-1/2 cups milk

3. Mix eggs and milk. Pour over bread. Refrigerate overnight.

1 (10-1/2 oz.) can cream of mushroom soup
1 cup (4 oz. pkg.) shredded sharp Cheddar cheese

4. Top with soup and cheese. Cover, and bake in Radarange Oven 15 minutes, turning every 5 minutes.

5. Let stand, covered, 5 minutes before serving.

Braised Duckling Spanish Style

Succulent dining for 4.

1 cut-up 5 lb. duckling

1. Preheat 9-1/2-inch Amana Browning Skillet in Radarange Oven 4-1/2 minutes. Place half of duckling pieces in skillet, and cook 1 minute per side. Drain off fat and reserve.
2. Heat Amana Browning Skillet in Radarange Oven 2 minutes. Place remaining half of duckling pieces in skillet and cook 1 minute per side. Drain fat. Place duckling pieces on paper towels.

1/4 cup all-purpose flour
2 teaspoons paprika
1/2 cup dry sherry
1-3/4 cups chicken bouillon

3. Return 3 tablespoons reserved fat to skillet. Cook in Radarange Oven 45 seconds. Blend in flour and paprika. Stir until smooth. Gradually stir in sherry. Blend in chicken bouillon. Mix well with wire whip.

1 sliced onion
1 sliced tomato
1/3 cup chopped, stuffed olives
2 tablespoons minced parsley

4. Cook sauce from step 3 in Radarange Oven 3 minutes. Stir after 2 minutes. Add browned duckling. Top with onion. Cook, covered, in Radarange Oven 20 minutes. Turn skillet halfway through cooking time. Garnish with tomatoes, olives and parsley.

Roast Turkey

All-American meal for 12.

13 to 14 lb. young turkey
Salt

1. Sprinkle turkey cavity with salt.

2 apples
2 tablespoons poultry seasoning
1 tablespoon packaged chicken gravy mix

2. Quarter apples. Roll apples in poultry seasoning and sprinkle with packaged gravy mix. Place apples in turkey cavity.

2 tablespoons shortening
1 tablespoon paprika

3. Truss turkey. Rub all over with blended shortening and paprika. Place turkey on large glass or earthenware dish. Cook in Radarange Oven as follows:
   Breast up: 20-22 minutes
   First side: 20-22 minutes
   Second side: 20-22 minutes
   Back up: 20-22 minutes
4. Remove turkey from Radarange Oven. Insert meat thermometer in center of turkey breast. Let turkey stand until temperature reaches 180° to 185° .

   MICRO-TIP: Turkey drippings may be used to make gravy.

Speedy, Baked Chicken

Speedy but satisfying dish to serve 4.

1/4 cup butter
1 cut-up broiler-fryer
Salt
Pepper
Paprika

1. Preheat 9-1/2-inch Amana Browning Skillet 4-1/2 minutes in Radarange Oven.
2. Place butter in Amana Browning Skillet. Arrange broiler-fryer pieces in skillet so that large pieces are around edge, skin side down. Place wings in center. Season with salt, pepper and paprika.
3. Cook in Radarange Oven 7 minutes per pound, or until chicken is tender. Turn after first 45 seconds. Turn again after next 3-1/2 minutes.

MICRO-TIP: Gravy may be made from drippings, if desired.

Shake and Bake Chicken

Simple but snazzy dish for 6.

2 lbs. chicken pieces
1 pkg. seasoned coating mix

1. Wash chicken. Coat with seasoned mix by shaking in bag according to directions on package.
2. Arrange chicken in 2-quart utility dish with larger pieces, such as thighs and breasts, at corners. Place small pieces such as legs and wings at center.
3. Bake in Radarange Oven 16 minutes. Turn dish halfway through cooking time.

Chicken Supreme

Utmost in chicken cookery for 6.

3/4 teaspoon seasoned salt
Paprika
3 whole, split chicken breasts

1 chicken bouillon cube
1/4 cup hot water
1 teaspoon instant minced onion
1/2 teaspoon curry powder
1/2 cup Sauterne

1 (3 oz.) can sliced, drained mushrooms

1. Sprinkle seasoned salt and paprika generously over chicken breasts. Arrange chicken in 10-inch ceramic skillet. Cook, uncovered, in Radarange Oven about 10 minutes.
2. Combine bouillon cube, water, onion, curry powder and Sauterne. Pour over chicken. Cook in Radarange Oven 15 to 20 minutes. Turn chicken every 3 to 4 minutes.
3. Stir in mushrooms 3 minutes before cooking time ends.

MICRO-TIP: If desired, chicken may be browned in Amana Browning Skillet.

Chicken Wings Mediterranean

A Mediterranean marvel for 6.

2-1/2 lbs. chicken wings

1. Cut tips from wings.

1/3 cup all-purpose flour
1 teaspoon salt

2. Measure flour and salt into paper bag. Coat chicken. Heat in Radarange Oven 5 minutes using 2-quart utility dish. Turn chicken halfway through cooking time.

2 cups sliced carrots
1/2 cup chopped onion
1 tablespoon parsley flakes
2 whole cloves
1 teaspoon garlic salt
1/4 teaspoon basil

3. Arrange carrots and onion around chicken. Sprinkle spices on top of chicken and vegetables.

1-3/4 cups tomato juice
1 tablespoon lemon juice

4. Combine lemon juice with tomato juice. Pour over chicken. Cook, covered, in Radarange Oven 18 to 20 minutes. Turn dish every 4 minutes. Chicken and carrots should be just tender.

MICRO-TIP: Serve with cooked rice or noodles.

Chicken Tamale Pie

Chicken today! Tamale tonight! 6 to 8 servings.

1/2 cup chopped onion
1/2 cup chopped celery
1 crushed clove garlic
3 tablespoons vegetable oil

1. Sauté onion, celery and garlic in 2 tablespoons oil in 10-inch ceramic skillet. Place in Radarange Oven 3 minutes. Stir halfway through cooking time.

2-1/2 cups chicken broth
1-1/2 cups diced, cooked chicken
1 (8 oz.) can tomato sauce
1 (2-1/4 oz.) can chopped ripe olives
2 teaspoons chili powder
2 teaspoons salt

2. Stir in 1/2 cup chicken broth, chicken, tomato sauce, olives, chili powder and 1 teaspoon of salt. Bring remaining broth to boil. Mix in remaining salt and oil.

1 cup yellow cornmeal
1 cup cold water

3. Combine cornmeal and water. Gradually stir into boiling broth. Cook in Radarange Oven 5 minutes. Stir every 2 minutes. Pour half of mixture into greased 2-quart casserole. Cover with chicken mixture. Top with remaining cornmeal.

Paprika

4. Sprinkle top with paprika. Bake in Radarange Oven 15 minutes. Turn dish halfway through cooking time.

Chicken Livers Chablis

8-10 pleasing portions.

2 tablespoons butter

1. Melt butter in 10-inch ceramic skillet in Radarange Oven for 1 minute.

16-20 (1-1/2 lbs.) chicken livers
Salt
Pepper

2. Season chicken livers with salt and pepper.

2-1/2 tablespoons all-purpose flour

3. Dredge livers in flour. Arrange in melted butter. Cook, uncovered, in Radarange Oven 7 minutes. Turn livers over halfway through cooking time.

3/4 cup Chablis (or other dry white wine)
1/4 cup minced onion
2 tablespoons catsup

4. Lightly stir in wine, onion and catsup. Cook in Radarange Oven 3 minutes. Serve with hot rice.

Chicken Liver Kabobs

Kabobs for 4.

8 large chicken livers (about 12 oz.)

1. Halve chicken livers with scissors, and snip out any veiny parts or skin.
2. Prepare 10-inch ceramic skillet by lining with double layer of paper towels.

2 slices bacon (divided in 8 small pieces)
1 (8 oz.) can small, well drained, white onions

3. Arrange kabobs on 4 wooden skewers in the following order: a whole onion, bacon, chicken liver. Repeat 3 times. Begin and end with onion.

2 tablespoons Worcestershire sauce

4. Brush kabobs with Worcestershire sauce. Place in skillet and cover with paper towel. Cook in Radarange Oven 8 to 10 minutes, or until bacon and livers are nicely browned.

MICRO-TIP: Wooden skewers can be purchased in novelty departments and import shops.

Roasted Pheasant

Pleasant pheasant for 4.

1 pheasant
1 tablespoon dry tarragon
3 tablespoons unsalted, sweet butter
2 tablespoons olive oil
1/2 teaspoon dry hot red pepper flakes

1. Wash cavity of bird. Spoon tarragon and 1 teaspoon butter into cavity, then truss bird. Rub bird with 1 tablespoon olive oil, then roll in hot pepper.

2 cloves garlic

2. Preheat 9-1/2-inch Amana Browning Skillet in Radarange Oven 4-1/2 minutes. Melt remaining butter and oil in skillet in Radarange Oven. Add garlic. Sauté until brown. Remove garlic and place pheasant in skillet. Brown pheasant.

Salt

3. Remove bird from skillet and place on glass platter. Cover, and cook in Radarange Oven 20 to 25 minutes. Baste every 5 minutes with pheasant's natural juices. Turn pheasant halfway through cooking time. Let stand 10 minutes before serving. Salt lightly.

Quail On Rice Nest

Quality quail for 4.

4 quail
3 tablespoons olive oil
1/4 teaspoon paprika
2 tablespoons unsalted, sweet butter

1. Clean quail. Reserve livers. Truss. Rub quail well with 2 tablespoons olive oil and paprika. Preheat 9-1/2-inch Amana Browning Skillet in Radarange Oven 4-1/2 minutes. Melt 1 tablespoon butter and 1-1/2 teaspoons olive oil in Radarange Oven in skillet. Place 2 quail in skillet. Brown. Preheat skillet again in Radarange Oven 2-1/4 minutes. Place remaining butter and olive oil in skillet. Brown remaining quail. Remove quail and keep warm.

1 cup Madeira wine
1 cup hot chicken broth
1/4 teaspoon salt

2. Pour fat from skillet, leaving browned particles. Stir in wine. Blend in chicken broth and salt. Place this sauce in glass bowl. Cook in Radarange Oven 5 minutes. Stir every minute. Reduce the liquid by about half. Remove from Radarange Oven.
3. Chop quail livers. Stir into glass bowl with sauce. Add quail. Cover. Cook in Radarange Oven 4 minutes. Turn quail halfway through cooking time. Baste. Remove from Radarange Oven and let stand, covered, 10 minutes.

2 cups long grain rice

4. Cook rice. Let stand, covered, 10 minutes. Arrange equal portions to resemble shallow nests in ceramic ramekins. Place 1 quail on center of each rice nest. Spoon sauce from glass bowl over quail. Cover birds in ramekins with plastic film wrap. Bake in Radarange Oven 10 minutes. Turn dishes halfway through cooking time. Let stand covered, 10 minutes, before serving.

Cinnamon
Nutmeg
Clove

81
Meats

Whether it's Cantonese Ribs, Veal Scallopini or Country Style Lamb Chops you desire, you'll find they're each delicious when prepared in the Radarange Oven. This is just a sampling of recipes found in this chapter. You can make extra servings, since leftover meats reheat very well in the Radarange Oven.

Meat items over 3 pounds will brown naturally in the Radarange Oven. You may brush a paste of water and dry onion soup mix over a roast to enhance flavor and color. You may also glaze a pork or lamb roast with your favorite jelly. An Amana Browning Skillet may also be used when browning is desired.

Remember to salt all meats after the baking time.

When cooking meats in the Radarange Oven, remember that tender meats cook very well and quickly with microwaves. Less tender cuts of meat need a long, slow simmering to become tender. You may use the Automatic Defrost or Slo Cook Cycle for this purpose. To tenderize a lesser cut of meat, brush the meat with soy sauce, or marinate the meat for 24 hours in olive oil, using an aromatic herb. Some herbs you might wish to try are marjoram, thyme, laurel, oregano and freshly ground black pepper. If the meat is a "white" meat such as veal or pork, add some lemon juice to the oil, using a few drops per 1/2 cup. For a red meat like beef, substitute a few drops of vinegar for the lemon juice.

Roasts

1. Place the roast, fat side down, on a cooking grill in a 2-quart utility dish. Season as desired. (It is best to salt the meat after cooking.) Cover the roast with waxed paper or a paper towel.
2. Turn the roast over halfway through the cooking time.
3. As moisture accumulates in the dish, baste off the liquid. Microwave energy is attracted by moisture and excess liquid may lengthen the cooking time.
4. Cook roasts as directed in the timetable. Allow 15 to 20 minutes before checking the desired temperature, using a meat thermometer.

KIND OF ROAST	TIME PER POUND	TEMPERATURE AFTER STANDING TIME
Beef	Rare—6-1/2 min. Medium—7-1/2 min. Well-done—8-1/2-9 min.	140° 160° 170°
Veal	Medium—8-8-1/2 min. Well-done—9-9-1/2 min.	160° 170°
Pork	Well-done—10 min.	170°
Ham (Pre-cooked)	5 min.	150°
Lamb	Well-done—9-1/2 min.	180°

WIENERS, SAUSAGE AND BACON

Bacon

1. Arrange bacon slices on a cooking grill in a 2-quart utility dish. Cover slices with a paper towel.
2. Cook the bacon in the Radarange Oven until it is browned and crisp.

2 slices	2 minutes
4 slices	4 minutes
12 slices	6-7 minutes

Link Sausages

1. Arrange 8 ounces of link sausages on a cooking grill in a 2-quart utility dish. Cover the sausages with a paper towel.
2. Cook in the Radarange Oven 5 minutes, or until the sausages are done.

MICRO-TIP: Sausages may also be cooked in an Amana Browning Skillet.

Wieners

1. Place wieners on a cooking grill in a 2-quart utility dish. Cover the wieners with a paper towel.
2. Cook in the Radarange Oven until steaming hot:

2 wieners	1 minute
4 wieners	2 minutes
6 wieners	2-1/2-3 minutes

Spiced Cider Baked Ham

Beef and Snow Peas Casserole

A touch of Chinese cooking for 4-6.

- 1 (7 oz.) pkg. frozen Chinese pea pods
- 1 lb. ground beef
- 1/2 cup green onions
- 1 (10-3/4 oz.) can cream of mushroom soup
- 1 tablespoon soy sauce
- 1/8 teaspoon pepper
- 1 (3 oz.) can chow mein noodles

1. Place package of frozen pea pods in Radarange Oven for 1-1/2 minutes to partially thaw.
2. Crumble meat into 1-1/2-quart casserole. Slice onions, diagonally. Mix with meat.
3. Layer peas over meat mixture.
4. Blend soup, soy sauce, and pepper. Spoon over peas. Bake in Radarange Oven 14 minutes until meat mixture loses its pink color. Turn dish every 4 minutes. Let stand 5 minutes before serving.
5. Heat noodles on paper plate in Radarange Oven 1-1/2 minutes. Turn dish halfway through cooking time. Sprinkle over casserole.

Rice And Spice Hamburger Casserole

6 hearty servings.

- 1 lb. ground beef
- 1/2 cup chopped onion
- 1 (10-3/4 oz.) can golden mushroom soup
- 1 (10-3/4 oz.) soup can water
- 1 cup thinly sliced carrot
- 1/2 cup chopped celery
- 1 chicken bouillon cube
- 2 tablespoons soy sauce
- 1/4 teaspoon ground thyme
- 1 cup pre-cooked rice
- 3/4 cup water

1. Brown beef and onion in covered ceramic baking dish in Radarange Oven for 6 minutes. Stir halfway through cooking time.
2. Stir in all ingredients but rice and water. Cook in Radarange Oven 10 minutes. Stir halfway through cooking time.
3. Moisten rice with water. Stir lightly into meat mixture. Cover, and cook in Radarange Oven 10 minutes. Stir halfway through cooking time. Let stand 5 to 10 minutes before serving.

Beef 'N Tater

Casserole for 6 sure compliments.

- 1 lb. ground beef
- 1 (1 lb.) pkg. frozen Tater Tots
- 2 teaspoons instant minced onion
- 1 (10-3/4 oz.) can cream of celery soup
- 1 (10-3/4 oz.) can golden mushroom soup

1. Crumble ground beef in 2-quart casserole. Cook in Radarange Oven 6 minutes. Turn dish halfway through cooking time. Drain.
2. Top with Tater Tots and onion.
3. Mix soups together. Pour on top of Tater Tots. Bake in Radarange Oven 8 to 10 minutes. Turn dish halfway through cooking time.

Ground Beef Stroganoff

Economical, but tasty stroganoff for 4.

1 lb. ground beef
1/2 cup finely chopped onion

1. Cook ground beef and onion in a 10-inch ceramic skillet in Radarange Oven for 6 minutes. Stir every 2 minutes.

1 (4 oz.) can drained mushrooms
2 tablespoons all-purpose flour
1 teaspoon salt
1/2 teaspoon paprika

2. Add mushrooms, flour, salt and paprika.

1 (10-1/2 oz.) can condensed cream of chicken soup
1 (8 oz.) carton dairy sour cream

3. Stir in soup. Cook in Radarange Oven 5 minutes, stirring halfway through cooking time. Add sour cream. Return to Radarange Oven for 2 minutes. Turn dish halfway through cooking time.

MICRO-TIP: Serve over cooked noodles or rice.

California Pilaff

Serve at your table of 4!

1 lb. ground beef

1. Sauté meat in ceramic skillet in Radarange Oven for 7 minutes. Stir after every minute.

2 cups hot water
1 (6 oz.) can tomato paste
1/2 cup uncooked long grain rice
1/3 cup finely chopped green pepper
1/3 cup finely chopped onion
1 small minced clove garlic
1-1/2 teaspoons salt
1/4 teaspoon pepper

2. Add remaining ingredients. Cover, and cook in Radarange Oven 25 minutes. Stir every 8 minutes.

Quick Chow Mein

A quick meal to present to 4-6 family members.

1 lb. ground beef
1/4 cup chopped onion

1. Crumble ground beef with onion in ceramic skillet. Cover, and cook in Radarange Oven 6 to 7 minutes. Stir halfway through cooking time.

1 (10-1/2 oz.) can cream of mushroom soup
1 (16 oz.) can Chinese vegetables
1 teaspoon soy sauce
1 teaspoon salt

2. Add soup, vegetables and seasonings. Stir well. Cook in Radarange Oven 5 minutes. Stir halfway through cooking time.

MICRO-TIP: Serve over hot cooked rice or with chow mein noodles.

Polynesian Medley

South Sea medley for 8.

2 tablespoons butter or margarine
1 lb. halved chicken livers

1. Melt butter in 10-inch ceramic baking dish in Radarange Oven 1 minute. Stir in livers. Cover and sauté 6 minutes. Stir every 2 minutes. Remove livers from baking dish. Place in center of 3-quart glass casserole.

1 lb. ground beef
1 egg
1/4 cup bread crumbs
2 tablespoons milk
1 teaspoon salt
1/8 teaspoon nutmeg

2. Combine ground beef, egg, bread crumbs, milk, salt and nutmeg. Mix lightly until well-blended. Shape into 1-inch balls.

3. In same ceramic baking dish, brown meat balls on all sides, about 5 minutes. Place meat balls around livers in 3 groups.

1 (16 oz.) can pineapple tidbits
1/2 lb. cocktail frankfurters

4. Drain pineapple and reserve juice. Arrange frankfurters and pineapple between meat balls.

1/4 cup firmly packed brown sugar
2 tablespoons cornstarch
1 teaspoon chicken bouillon powder
3 tablespoons vinegar
1 tablespoon soy sauce

5. Mix brown sugar, cornstarch, and bouillon in glass measure. Measure 3/4 cup pineapple juice. Pour juice, vinegar and soy sauce into dry ingredients. Mix well. Cook in original baking dish in Radarange Oven until thickened, about 5 minutes. Stir every 2 minutes. Pour over meats and pineapple. Cover and refrigerate until just before serving time.

6. Uncover and bake in Radarange Oven for 13-1/2 minutes. Turn dish every 5 minutes.

Stuffed Peppers

4 zesty peppers.

4 medium green peppers

1. Wash peppers. Remove seeds and membrane. Parboil in salted boiling water in Radarange Oven 4-5 minutes. Drain.

1 cup finely diced, cooked roast beef
1 cup cooked rice
1 (7-1/2 oz.) can tomato sauce
1/4 cup finely chopped celery
1 tablespoon minced onion
1/2 teaspoon seasoned salt

2. Combine remaining ingredients and fill peppers. Place peppers in 9-inch dish. Bake in Radarange Oven 10-11 minutes.

MICRO-TIP: Prepare peppers ahead, then refrigerate or freeze. Remember to allow extra cooking time.

Chinese Pepper Beef

Choice Chinese dish for 4.

1 cup hot beef broth

1. Heat beef broth in Radarange Oven 1-1/2 minutes. Set aside.

2 tablespoons vegetable oil
1 lb. thinly-sliced beef round steak
1 minced clove of garlic

2. Preheat 9-1/2-inch Amana Browning Skillet in Radarange Oven 4-1/2 minutes. Place 1 tablespoon oil, half of beef and half of minced garlic in skillet. Cook in Radarange Oven 1 minute. Stir halfway through cooking time.
3. Wipe skillet with paper towel. Preheat skillet in Radarange Oven 2 minutes. Cook remaining beef and garlic in oil in Radarange Oven 1 minute. Stir halfway through cooking time.
4. Place cooked beef and beef broth in skillet. Heat 1 minute in Radarange Oven.

2 tablespoons cornstarch
4 teaspoons soy sauce
2 tablespoons water
1 teaspoon finely minced ginger
1 finely sliced, large green pepper

5. Mix cornstarch, soy sauce and water to smooth paste. Stir into skillet. Blend in ginger and green pepper slices. Cook in Radarange Oven 3 minutes. Stir halfway through cooking time.

Swedish Meat Balls

Scandinavian special — 1 dozen meat balls.

1 cup milk
1/2 cup bread crumbs

1. Pour milk over bread crumbs. Let stand 15 minutes.

1/2 lb. ground round steak
1/2 lb. ground lean pork
1 egg
1 teaspoon salt
1/4 teaspoon pepper
1 tablespoon steak sauce
1 tablespoon instant minced onion

2. Lightly mix in meats, then egg and seasoning. Shape into 12 balls.

2 tablespoons shortening

3. Preheat 9-1/2-inch Amana Browning Skillet 4-1/2 minutes. Melt shortening in skillet. Brown meat balls in shortening, turning them to brown all sides. Remove meat balls with slotted spoon and reserve.

1/4 cup all-purpose flour
1 cup water
2 teaspoons instant bouillon

4. Sprinkle flour into hot shortening in skillet. Stir until smooth. Gradually blend in water and bouillon, stirring constantly. Heat in Radarange Oven 3 minutes. Stir every 30 seconds.
5. Place meat balls in gravy. Cover, and cook 3 additional minutes in Radarange Oven.

Salisbury Steak

6 succulent servings of steak.

1-1/2 lbs. ground beef
1 (10-1/2 oz.) can golden mushroom soup
1/2 cup bread crumbs
1 slightly-beaten egg
1/4 cup chopped onion
Dash of pepper

1. Combine ground beef, 1/4 cup soup, bread crumbs, egg, onion and pepper. Shape into 6 patties. Using 2-quart utility dish, cook in Radarange Oven 10 minutes. Turn patties and drain fat halfway through cooking time.

1/3 cup water

2. Blend remaining soup with water. Pour over meat. Bake in Radarange Oven 4 minutes. Turn dish halfway through cooking time

Meat Loaf

Meat loaf for 4-6.

1 clove garlic

1. Rub 1-1/2-quart utility dish with garlic.

1 lb. ground beef
6 crushed saltine crackers
1/4 cup minced onion
1/4 cup chopped green pepper
1/4 cup chopped celery
1 beaten egg
1 tablespoon minced parsley
1/2 cup tomato juice
3/4 teaspoon salt
1/4 teaspoon pepper

2. Combine all remaining ingredients except catsup, mixing lightly but thoroughly. Spoon mixture evenly into utility dish. Cook in Radarange Oven 6 minutes, turning dish every 2 minutes.

1/4 cup catsup

3. Drizzle catsup over top of loaf. Cook in Radarange Oven 5 minutes.

Steak Cups

6 cups of good taste.

1 lb. ground sirloin
1/2 cup milk
1 slightly-beaten egg
1/2 cup quick-cooking rolled oats
1/3 cup finely chopped green pepper
2 teaspoons prepared horseradish
2 teaspoons prepared mustard
1/2 teaspoon salt

1. Combine all ingredients except catsup, mixing lightly but thoroughly. Divide into 6 equal portions. Place in 5-ounce greased custard cups. Make hollow in center of each portion.

6 tablespoons catsup

2. Fill hollows with catsup. Cook in Radarange Oven 10 minutes.

Shish Kabobs

Kabobs for 2!

16 (3/4-in.) cubes sirloin steak
16 squares green pepper
12 fresh or canned mushroom caps
8 small canned or frozen whole onions

1. Thread meat and vegetables alternately on wooden skewers starting and ending with steak cubes. Lay four kabobs across 1-1/2-quart utility dish.

Steak baste or liquid marinade

2. Brush with liquid on all sides. Cook 5 to 7 minutes until meat is cooked to desired degree. Roll kabobs over halfway through cooking time.

MICRO-TIP: Double recipe for 4 servings, allow 2 shish kabobs per person.

Peppered Tenderloin

A mouth-watering delight serving 8 tenderloin lovers.

2 lbs. boneless beef tenderloin or sirloin

1. Slice beef into 1/4-inch wide strips. Preheat 9-1/2-inch Amana Browning Skillet 4-1/2 minutes in Radarange Oven.

4 tablespoons butter or margarine

2. Place butter in skillet. Brown the meat in butter. Then sauté half of meat in butter 4 minutes. Stir halfway through cooking time.

1 teaspoon salt
1/2 teaspoon coarsely ground pepper
Dash of ground sage
Dash of ground cumin

3. Remove meat with slotted spoon to 3-quart casserole. Sauté other half of meat in same way. Sprinkle beef with salt, pepper, sage and cumin. Toss lightly to mix.

1 lb. fresh sliced mushrooms
1 medium onion
2 medium green peppers
2 minced cloves garlic
2 medium tomatoes

4. Sauté mushrooms in Amana Browning Skillet in Radarange Oven about 2 minutes. Stir into meat. Cut onion into eighths. Cut peppers into 1-inch pieces. Blend garlic, onion and green pepper into drippings in Amana Browning Skillet. Sauté 2 minutes. Stir into meat. Cut each tomato into 8 wedges. Stir into meat.

1/2 cup soy sauce
2 tablespoons vinegar
2 tablespoons tomato paste

5. Combine soy sauce, vinegar and tomato paste in Amana Browning Skillet. Heat in Radarange Oven 2 minutes. Stir halfway through cooking time. Pour over meat. Toss lightly to mix. Cover.

6. Cook casserole in Radarange Oven 15 minutes, or until vegetables barely tender. Stir every 5 minutes. Let stand few minutes before serving.

Sukiyaki

Succulent Sukiyaki for 4.

1 lb. round steak

1. Slice steak very thinly, diagonally across the grain.

1/2 lb. fresh bamboo shoots
1/4 lb. fresh water chestnuts
1 lb. fresh bean sprouts
1/2 lb. fresh mushrooms
1 medium onion
3 stalks celery

2. Drain bamboo shoots and water chestnuts. Slice thinly. Drain bean sprouts and rinse with cold water. Clean mushrooms and onion. Slice thinly. Slice celery diagonally.

1 tablespoon shortening

3. Preheat 9-1/2-inch Amana Browning Skillet in Radarange Oven 4-1/2 minutes. Melt shortening in skillet. Brown beef strips.

4. Arrange browned meat in center of skillet. Distribute vegetables around the meat.

3 tablespoons sugar
1/3 cup soy sauce
1/2 cup bouillon

5. Combine sugar, soy sauce and bouillon. Pour over meat and vegetables.

6. Cook in Radarange Oven 4 to 5 minutes. Do not overcook. Vegetables should be quite crisp.

Beef Stroganoff

Add a European flavor to your meal tonight: 3-4 servings.

1 lb. 1/4-inch-thick beef tenderloin
3 tablespoons all-purpose flour
1-1/2 teaspoons salt
1/4 teaspoon pepper
1 clove of garlic

1. Trim any excess fat from meat. Combine flour, salt and pepper. Rub both sides of meat with clove of garlic. Pound flour mixture into both sides of meat.

1/4 cup salad oil

2. Cut beef into strips 1-1/2-inch x 1-inch. Preheat 9-1/2-inch Amana Browning Skillet in Radarange Oven 4-1/2 minutes. Place salad oil and meat in skillet. Brown meat on all sides.

1/3 cup minced onion
1/4 cup water
1 (10-3/4 oz.) can cream of chicken soup
1 (4 oz.) can sliced mushrooms

3. Stir in onion. Cook in Radarange Oven 3 minutes. Stir in water, soup and mushrooms. Cook 4 to 5 additional minutes until beef just tender.

1 cup commercial sour cream

4. Stir sour cream in gently. Heat one additional minute before serving.

Cantonese Ribs

Radarange Swiss Steak

Steak enough for 6.

3 lbs. 2-inch thick rump or round steak
Half clove garlic
1/3 to 1/2 cup all-purpose flour

1. Rub meat with garlic, and then pound flour into it.

3 to 4 tablespoons suet or bacon fat

2. Heat suet or bacon fat in 3-quart casserole or ceramic Dutch oven. Place meat in dish. Heat in Radarange Oven 20 minutes. Turn meat over halfway through cooking time.

2 cups canned tomatoes
1 cup uniformly chopped carrots
1 cup uniformly chopped celery
1/2 cup chopped onion
2 teaspoons salt

3. Place vegetables around edge of casserole dish. Salt. Cook, covered, in Radarange Oven 30 minutes. Turn meat over, and turn dish halfway through cooking time. Remove lid for last five minutes of cooking, but cover with paper towel to avoid splatters.

MICRO-TIP: To thicken sauce, mix 1 tablespoon cornstarch with small amount water. Add to gravy and cook in Radarange Oven 1 minute. Pour over steak.

Pot Roast with Vegetables

An all-American meal for 6.

3 lbs. pot roast
2 large sliced onions

1. Render suet trimmed from roast in 10-inch ceramic skillet by cooking in Radarange Oven 5 minutes. Discard suet. Place roast in fat. Lay onion slices over roast. Cover, and cook 10 minutes in Radarange Oven.

1/2 cup beef stock or
1/2 cup red wine

2. Turn roast over. Add beef stock or wine. Cover, and cook 30 minutes, turning dish and roast halfway through cooking time.

4 peeled, sliced medium carrots
6 scraped small red potatoes
2 stalks diagonally-cut celery

3. Arrange vegetables around roast. Cook in Radarange Oven 15 minutes. Turn dish halfway through cooking time. Drain. Serve.

Braised Lamb Chops

Lamb for a lark: 4 generous servings.

4 (1-1/2-inch thick, each) lamb steaks or lamb shoulder chops
2 teaspoons grated lemon peel
1 teaspoon oregano
1/4 teaspoon garlic powder
1/8 teaspoon pepper

1. Combine lemon, oregano, garlic and pepper. Rub onto meat surfaces.

4 small red potatoes

2. Scrub, then pierce potatoes. Place in circle in Radarange Oven. Partially cook potatoes for 5 minutes.

1 tablespoon butter or margarine
1 large, thinly sliced onion

3. Melt butter in 2-quart utility dish 2 minutes in Radarange Oven. Stir in onion. Cook 2 minutes until translucent. Layer chops in dish. Spoon onion around meat. Cook in Radarange Oven 15 minutes. Turn meat and rotate dish after 10 minutes.

4 medium sized sliced carrots
2 tablespoons dry Vermouth
Salt

4. Arrange potatoes and carrots around meat. Sprinkle with wine. Cook in Radarange Oven, covered, for 7 minutes until carrots are tender. Season to taste.

1 teaspoon cornstarch
Chopped parsley

5. Pour dish juices into 2-cup glass measure. Stir in cornstarch. Cook in Radarange Oven 3 minutes. Pour over meat and vegetables. Sprinkle with parsley.

Liver Bake

Luscious liver for 6.

1/4 cup all-purpose flour
1 teaspoon salt
Dash pepper

1. Mix flour, salt and pepper in plastic bag. Use to coat liver.

6 slices bacon
1 cup finely chopped onion
1 lb. beef liver

2. Cook bacon and onion in 10-inch ceramic skillet 8 minutes in Radarange Oven. Remove bacon and onions. Reserve bacon drippings. Cook liver in skillet 8 minutes. Turn dish and turn liver halfway through cooking time.

1-1/2 cups milk

3. Place cooked liver in 1-1/2-quart casserole. Place remaining flour mixture in bacon fat. Gradually pour in milk. Cook in Radarange Oven 5 minutes, or until thickened. Stir every 2 minutes. Pour over liver.

1/4 cup dry bread crumbs
1 tablespoon melted butter
1/8 teaspoon paprika

4. Combine bread crumbs, butter and paprika. Sprinkle over top of casserole. Bake in Radarange Oven 5 minutes. Turn dish halfway through baking time.

Chinese Pork and Rice

4 Far-Eastern servings.

2 tablespoons vegetable oil
2/3 cup uncooked rice

1. Preheat 9-1/2-inch Amana Browning Skillet in Radarange Oven 2 minutes. Sauté rice in oil 3 minutes in Radarange Oven.

1-1/2 cups boiling water
1 bouillon cube
2 teaspoons soy sauce
1 teaspoon salt

2. Add water, bouillon cube, soy sauce and salt. Cover and cook in Radarange Oven 5 minutes.

1 cup diced, cooked pork
1 medium chopped green pepper
1/2 cup chopped onion
2 stalks chopped celery

3. Gently mix vegetables and pork throughout rice mixture. Cook, covered, in Radarange Oven for 8 to 8-1/2 minutes.

Roast Pork and Sauerkraut

German dish for 4.

4 large (2 oz. each) slices cooked roast pork
Salt
Pepper

1. Sprinkle pork lightly with salt and pepper.

1 (1 lb. 4 oz.) can sauerkraut
1/2 cup sauerkraut juice
1 teaspoon caraway seed

2. Drain kraut. Reserve 1/2 cup juice. Turn kraut into 1-1/2-quart casserole. Add kraut juice and caraway seed. Arrange meat slices on top of kraut mixture. Heat in Radarange Oven 4 minutes.

Chop Suey

5-6 servings of sauce to be served over rice or noodles.

1 lb. cubed pork, beef or veal

1. Preheat 9-1/2-inch Amana Browning Skillet in Radarange Oven 4-1/2 minutes. Brown meat in skillet in Radarange Oven 5 to 6 minutes, or until meat loses its pink color. Stir halfway through cooking time.

1/2 teaspoon salt
1/2 cup water

2. Sprinkle meat lightly with salt. Pour water over meat. Cook, covered, in Radarange Oven 3 minutes.

1 (1-5/8 oz.) pkg. chop suey sauce mix
1 (16 oz.) can Chinese vegetables

3. Stir in sauce mix and vegetables including vegetable liquid. Bring to boil. Cook in Radarange Oven 10 minutes.

MICRO-TIP: If you prefer thicker sauce, mix 1 teaspoon cornstarch with small amount water. Add during last two minutes of cooking.

Cantonese Ribs

Charming Chinese dish for 6.

4 lbs. country style pork ribs

1. Separate ribs. Place in 2-quart utility dish.

1/2 cup soy sauce
1/2 cup dry sherry
1/2 cup lemon juice
1-1/2 tablespoons brown sugar
1 teaspoon grated lemon peel
1/2 teaspoon garlic powder
1/4 teaspoon ginger

2. Combine soy sauce, sherry, lemon juice and seasonings in 2-cup glass measure. Pour over ribs. Cover with plastic wrap. Let stand at room temperature 2 hours. Baste frequently. Drain, reserving 1/2 cup marinade.

3. Bake ribs in Radarange Oven 38 minutes. Turn ribs and dish every 10 minutes. Drain off excess juices throughout cooking time.

1 cup orange marmalade

4. Mix marmalade and remaining marinade. Baste ribs. Cook in Radarange Oven 8 minutes.

Tangy Brown Ribs

4 spicy servings.

2 lbs. pork ribs
Salt
Pepper

1. Place ribs in 2-quart utility dish. Sprinkle lightly with salt and pepper. Bake in Radarange Oven 25 minutes. Turn dish every 5 minutes. Drain fat.

1 cup chopped onion
2 tablespoons brown sugar
2 tablespoons vinegar
1 teaspoon prepared mustard

2. Combine onion, brown sugar, vinegar and mustard. Spread evenly over ribs.

1 (15 oz.) can tomato sauce

3. Drench ribs with tomato sauce. Cook in Radarange Oven 10 minutes. Turn ribs and dish halfway through cooking time.

Sweet-Sour Pork

4 sweet-sour servings.

1 lb. cubed boneless pork loin

1. Preheat 9-1/2-inch Amana Browning Skillet in Radarange Oven 4-1/2 minutes. Brown meat in skillet 8 minutes. Turn meat every 2 minutes.

1 (2 oz.) pkg. sweet-sour sauce mix
3/4 cup water
1/2 cup pineapple juice
1/3 cup pineapple tidbits

2. Mix remaining ingredients in 2-cup glass measure. Heat 2 minutes in Radarange Oven. Pour over meat, covering thoroughly. Cook 4 minutes, or until sauce thickens. Turn halfway through cooking time.

London Broil

4 to 6 thinly sliced servings.

1 to 1-1/2 lbs. flank steak

2 tablespoons sherry
2 tablespoons soy sauce
1 tablespoon honey
1 minced clove garlic
1 tablespoon sugar
1 teaspoon salt
1/8 teaspoon pepper

2 tablespoons salad oil

1. Pierce surface of steak with sharp fork.
2. Combine sherry, soy sauce, honey, garlic, sugar, salt and pepper. Spread over steak. Refrigerate 10 to 12 hours. Baste occasionally.
3. Preheat 9-1/2-inch Amana Browning Skillet 4-1/2 minutes in Radarange Oven. Using 1 tablespoon oil, cook half of steak in skillet 3 to 4 minutes. Turn halfway through cooking time. Preheat skillet again for 2 minutes. Repeat cooking process with remaining oil and steak.

Quick Cassoulet

Luscious lamb for a lucky family of 6.

1 cup chopped onion
1/2 lb. halved sausage links
1 minced clove garlic

1-1/2 cups boiling water
1 tablespoon minced parsley
1/2 teaspoon crushed thyme
2 cups cubed, cooked lamb

2 (16 oz. each) cans Boston style baked beans

1 (8 oz.) can tomato sauce

1. Preheat 9-1/2-inch Amana Browning Skillet in Radarange Oven 4-1/2 minutes. Brown onion, sausage and garlic in skillet 3 minutes. Stir halfway through cooking time.
2. Stir in water and seasonings. Cook in Radarange Oven 3 minutes. Drain. Reserve liquid. Mix in lamb.
3. Spread layer of beans in 3-quart casserole, followed by layer of sausage, lamb and onion. Top with layer of baked beans.
4. Pour tomato sauce into reserved liquid and mix. Pour over beans. Cook, covered, in Radarange Oven 12 minutes. Turn dish halfway through cooking time.

Quick Cassoulet

Spiced Cider Baked Ham

4 servings of cider sauce and ham to serve special people!

2 (1-inch thick) center cut slices of ham steak

1. Preheat 9-1/2-inch Amana Browning Skillet in Radarange Oven 4-1/2 minutes. Cook one steak 45 seconds per side. Reserve drippings and reheat skillet 2-1/4 minutes. Repeat cooking time with second steak.

1 cup sweet cider or apple juice
1 large sliced onion
3 tablespoons brown sugar

2. With both steaks in skillet, pour cider, onion and sugar over top. Cook in Radarange Oven 6 to 8 minutes. Turn dish halfway through cooking time. Prepare spiced cider sauce using 1/2 cup ham drippings.

1/2 cup ham drippings
1/2 cup corn syrup
1/2 cup seedless raisins

3. Combine drippings, corn syrup and raisins in small bowl. Cook in Radarange Oven 45 seconds.

2 tablespoons cornstarch
2 tablespoons water
1/4 teaspoon cinnamon
1/8 teaspoon nutmeg

4. Stir in cornstarch, water and spices. Cook in Radarange Oven 2 minutes until thickened. Stir halfway through cooking time. Serve sauce over ham.

Ham Tetrazzini

6-8 servings of a special preparation.

1 (4 oz.) can sliced mushrooms

1. Drain mushrooms. Reserve liquid. Add enough water to measure 2 cups.

1/2 cup chopped onion
1/2 cup chopped celery
6 tablespoons butter or margarine

2. Sauté onion and celery in butter using 10-inch ceramic skillet. Cook in Radarange Oven 4 minutes. Stir halfway through cooking time.

6 tablespoons all-purpose flour
1/4 teaspoon pepper

3. Mix in flour, pepper and reserved liquid. Stir until smooth. Cook 5 minutes.

1 cup light cream
2 chicken bouillon cubes

4. Blend in cream and dissolve bouillon cubes in mixture.

3 cups cubed ham
3 tablespoons dry sherry

5. Lightly stir in mushrooms, ham and sherry.

1 (7 oz.) pkg. cooked spaghetti
1/4 cup Parmesan cheese

6. Blend spaghetti into mixture. Sprinkle with Parmesan cheese. Bake in Radarange Oven 18 to 20 minutes, or until center of casserole is hot and slightly bubbly. Turn dish every 4 or 5 minutes.

Ham Potato Pea Scallop

A wallop of a scallop for 8!

1/4 cup chopped onion
1/4 cup butter or margarine

1. Sauté onion in melted butter in Radarange Oven 1-1/2 minutes in 1-quart casserole.

1/4 cup all-purpose flour
2 cups milk
3/4 teaspoon salt
1/8 teaspoon pepper

2. Blend in flour. Gradually stir in milk, salt and pepper. Cook in Radarange Oven 4-1/2 minutes. Stir well every 1-1/2 minutes.

6 medium cooked potatoes
1 (10 oz.) pkg. cooked green peas
2 cups diced, cooked ham
3 slices process American cheese

3. Combine potatoes, green peas and ham in 3-quart casserole. Cut cheese slices in half. Arrange over top. Cook in Radarange Oven 5 minutes. Turn dish halfway through cooking time. Let casserole stand 10 minutes before serving.

Radarange Bacon

Achin' for bacon? Try this!

Slice of bacon
1/2 lb. or 12 slices

Line paper plate with paper towel. Place bacon strip on plate. Cover with paper towel. Cook 45 to 60 seconds. Several slices may be arranged on cooking grill in 2-quart utility dish. Layer if necessary, using paper toweling between layers. Cook in Radarange Oven 45-60 seconds per slice depending on personal preference.

Baked Canadian Bacon Treat

Campers' delight, day or night: 8 portions.

2 lbs. sliced Canadian bacon

1. Preheat 9-1/2-inch Amana Browning Skillet in Radarange Oven 4-1/2 minutes. Brown bacon 45 seconds per slice.

1/2 cup firmly packed brown sugar
1/2 cup unsweetened pineapple juice
1/2 teaspoon dry mustard

2. Combine brown sugar, pineapple juice, and mustard. Pour over bacon. Cook in Radarange Oven 10 minutes. Baste every 3 minutes. Turn bacon halfway through cooking time.

Hollywood "Ham" Casserole

8 servings that rate applause.

2 lbs. cubed ham
1 medium head coarsely shredded cabbage
3 large cubed potatoes
3 large sliced carrots
3/4 cup diced onion
1/2 cup water
1 teaspoon salt
1/4 teaspoon pepper

1. Place all ingredients in 4-quart casserole. Cook, covered, in Radarange Oven 35 minutes. Stir every 10 to 12 minutes.

2. Let stand 15 minutes before serving.

Franks and Spaghetti

Frankly delicious! 6 to 8 servings.

6 slices bacon

1. Cook bacon in Radarange Oven 6 minutes. Reserve drippings.

1/2 cup chopped onion

2. Sauté onion in drippings. Cook in Radarange Oven 2 minutes.

1 (12 oz.) can whole kernel corn
1 (10-3/4 oz.) can tomato soup
1/2 cup chili sauce
1/2 teaspoon oregano
1/4 teaspoon basil
1 (5 oz.) pkg. cooked spaghetti
1 lb. sliced frankfurters

3. Combine corn, soup, onion, chili sauce, bacon and spices. Mix in spaghetti and franks. Heat in Radarange Oven 6 minutes. Stir halfway through cooking time.

3/4 cup shredded sharp Cheddar cheese

4. Sprinkle cheese over top of casserole. Melt 5 minutes in Radarange Oven.

Hearty German Supper

Oktoberfest for 6.

1 (16 oz.) can applesauce
1 (14 oz.) can drained, chopped sauerkraut
1/3 cup dry white wine
2 tablespoons brown sugar

1. Combine applesauce, sauerkraut, wine and brown sugar. Cook, covered, in Radarange Oven 5 minutes. Stir halfway through cooking time.

1 lb. small white potatoes
1 (16 oz.) can whole drained onions
1 (12 oz.) ring Polish sausage

2. Arrange scrubbed potatoes and onions around edge of dish. Place slashed sausage in dish center. Cook in Radarange Oven 8 to 10 minutes. Turn dish every 3 minutes.

Parsley

3. Let stand 5 minutes before serving. Sprinkle parsley over top.

Brown and Serve Sausage

Sausage for Sunday.

1 (8 oz.) pkg. fully cooked sausage

Method I:
Preheat 9-1/2-inch Amana Browning Skillet 4-1/2 minutes. Pierce sausage skin with prongs of fork. Place sausage in skillet. Cook in Radarange Oven 2 minutes. Turn sausage over halfway through cooking time.

2 links

Method II: Preheat 6-1/2-inch Amana Browning Skillet 2 minutes. Pierce sausage skin with prongs of fork. Place sausage in skillet. Cook in Radarange Oven 1-1/2 to 2 minutes. Turn sausage over halfway through cooking time.

Precooked Sausages

Sausage for snacking.

4 ozs. Bratwurst, Pepperoni, or other similar type of sausage

1. Pierce sausage skin with prongs of fork.

2. Preheat 6-1/2-inch Amana Browning Skillet 2 minutes. Place sausages in skillet. Cook, covered, in Radarange Oven 1 minute. Allow 5 minutes standing time in covered skillet.

Pork Chops Maui

An island treat for 4.

4 center cut, 1-inch thick pork chops

1. Preheat 9-1/2-inch Amana Browning Skillet in Radarange Oven 4-1/2 minutes. Place chops in skillet. Cook 5 minutes, turning chops halfway through cooking time. Drain fat.

1/2 cup canned, crushed, drained pineapple
1/4 cup chopped onion
1/4 cup firmly packed brown sugar
3 tablespoons cider vinegar
1 clove garlic
1 teaspoon salt
1/2 teaspoon grated orange peel
1/2 teaspoon ground ginger
1/4 teaspoon seasoned pepper
Dash Tabasco

2. Combine remaining ingredients in blender. Process until smooth. Pour over chops. Cook, covered, in Radarange Oven 7 minutes. Turn dish every 2 minutes. Let stand 5 minutes before serving.

Stuffed Pork Chops

4 sizzling, stuffed pork chops.

4 1-inch thick pork chops

1. Cut large gash or pocket into side of each chop.
2. Preheat 9-1/2-inch Amana Browning Skillet in Radarange Oven 4-1/2 minutes. Place chops in skillet. Cook 5 minutes. Turn halfway through cooking time. Drain fat.

1 cup bread crumbs
1 tablespoon instant minced onion
1 tablespoon melted butter
1 teaspoon parsley flakes
1/4 teaspoon salt
Dash pepper
Dash poultry seasoning

3. Mix bread crumbs, onion, butter and seasonings.

1 tablespoon hot water
4 slices apple
1/2 cup honey
Nutmeg

4. Gradually add hot water until stuffing is just moistened. Divide stuffing. Fill each chop. Return to skillet. Place apple slice on each chop. Brush generously with honey. Sprinkle with nutmeg.
5. Bake in Radarange Oven 7 minutes. Turn chops halfway through cooking time. Brush with honey.

Roast Leg of Lamb

10-12 portions of lamb you'll love.

9 lb. leg of lamb

1. Cook lamb on platter or on cooking grill over 2-quart utility dish. Allow 9 to 10 minutes per pound.

Garlic salt

2. Cook in Radarange Oven 25 minutes. Turn roast completely over. Cook 28 minutes. Sprinkle with garlic salt. Cook in Radarange Oven 14 minutes. Turn over. Sprinkle with garlic salt. Cook 14 minutes. Check temperature. Internal temperature should register 180° .

MICRO-TIP: Baste off juice as it accumulates in baking dish. Roast may be covered with paper towel to eliminate splattering.

Shepherd's Pie

Feed your flock of four this favorite.

2 cups cubed, cooked lamb
1 tablespoon all-purpose flour
2 tablespoons vegetable oil

1. Preheat 9-1/2-inch Amana Browning Skillet in Radarange Oven 4-1/2 minutes. Roll meat lightly in flour until coated. Brush skillet with oil. Brown meat lightly on all sides.

2 (7-3/4 oz. each) cans mushroom gravy
1 (8 oz.) can whole mushrooms
1 (8 oz.) can sliced carrots
1 (8 oz.) can green peas
Salt
Pepper

2. Stir in gravy, mushrooms, carrots and peas. Season with salt and pepper. Mix lightly to combine ingredients.

1-1/2 cups hot, mashed potatoes

3. Spoon potatoes on top. Heat in Radarange Oven 6 to 7 minutes.

Spanish Lamb Chops

Quatro (4) chops de Espanol.

4 (1/2 lb. each) shoulder lamb chops
1 tablespoon vegetable oil

1. Preheat 9-1/2-inch Amana Browning Skillet in Radarange Oven 4-1/2 minutes. Place oil, then chops in skillet. Cook in Radarange Oven 4 minutes. Turn meat over halfway through cooking time.

1 (16 oz.) can stewed tomatoes
1 sliced small green pepper
1/2 cup chopped onion
4 slices lemon
Salt
1 cup finely chopped cooked ham

2. Mix in tomatoes, green pepper, onion, lemon and salt. Cook, covered, in Radarange Oven 30 minutes using Slo Cook or Automatic Defrost control, or 12 minutes regular setting. Turn dish halfway through cooking time, mixing in ham.

Veal Scallopini

4 scallopini servings.

1 lb. thinly sliced veal
1/4 cup vegetable oil
1 clove garlic

1. Preheat 9-1/2-inch Amana Browning Skillet in Radarange Oven 4-1/2 minutes. Brown veal in oil and garlic 2 minutes. Turn meat over halfway through cooking time. Remove garlic and veal from skillet.

3/4 cup sliced onion
1 (6 oz.) can drained mushrooms
2 tablespoons all-purpose flour
1/2 teaspoon salt
1/8 teaspoon pepper

2. Stir onion and mushrooms into hot oil. Cook in Radarange Oven 3 minutes. Stir in flour, salt and pepper until well blended. Cook 2 minutes.

1 (8 oz.) can tomato sauce
1/2 cup water

3. Gradually stir in tomato sauce and water. Cook 5 minutes, stirring halfway through cooking time.

4. Arrange veal in skillet. Cook 5 to 6 minutes, or until veal is tender.

Veal Steak With Dandy Dumplings

Veal for variety:
6-8 servings.

2 tablespoons vegetable oil
2 lbs. veal steak
1/2 cup all-purpose flour
1 teaspoon paprika
1/4 teaspoon pepper

1. Preheat 9-1/2-inch Amana Browning Skillet in Radarange Oven 4-1/2 minutes. Cut veal steak into 2-inch pieces. Mix seasonings with flour. Use to coat veal pieces. Place 1 tablespoon oil and half of coated veal pieces in skillet. Cook in Radarange Oven 2 minutes. Turn once to brown all sides. Transfer meat to 2-1/2-quart utility dish. Drain skillet.

2. Repeat browning procedure with remaining veal pieces. Preheat skillet 2 minutes, then cook 2 minutes turning once. Transfer to 2-quart utility dish.

1 cup hot water or broth
1 (16 oz.) can small white onions
1 (10-1/2 oz.) can cream of chicken soup

3. Mix remaining ingredients. Add to meat. Cook, covered, in Radarange Oven 5 minutes. Stir and top with dumplings from following recipe.

1 cup all-purpose flour
2 teaspoons baking powder
1/4 teaspoon salt

4. Sift flour with baking powder and salt.

1/2 teaspoon poultry seasoning
1/2 teaspoon instant minced onion
1/4 teaspoon crushed celery seed
1/2 cup milk
2 tablespoons vegetable oil

5. Stir in seasonings. Mix in milk and oil.

1/4 cup melted butter or margarine
1/2 cup fine packaged dry crumbs

6. Melt butter in small bowl. Drop batter by tablespoons into butter, then crumbs, until coated. Arrange on top of veal steak casserole. Cook in Radarange Oven, uncovered, 5 minutes. Cover. Let stand 5 minutes before serving.

103
Eggs & Cheese

Les Oeufs (eggs) can be prepared in endless ways. In this chapter we'll show you how the Radarange Oven can make fluffy scrambled eggs in seconds. Breakfast will never be dull again after using this variety of scrambled egg recipes. Try Chinese Scramble, California Eggs or Scrambled Eggs Foo Yung for a change of pace. For brunch, serve our Onion Pie or Crustless Quiche Lorraine.

A quiche (pronounced kēēsh) is a savory, rather than sweet, custard pie. It is served as an hors d'oeuvre in a French meal, but it is accepted here as a brunch or supper dish. If you are proud of your custard-baking ability, this is a great way to "show it off". Quiche Lorraine has become accepted as a quiche flavored with Swiss cheese and crisp bacon. Serve it with a green salad and a glass of your favorite white wine.

About cooking eggs in the Radarange Microwave Oven:

1. Do not cook eggs in the shell. The rapid heat generated through microwave cooking expands the air inside the shell and causes it to burst.
2. Overcooking will cause a tough and rubbery egg.
3. When preparing poached eggs, always be sure they are completely covered with water.
4. Before frying or poaching eggs, gently puncture the yolk with the tines of a fork. This will break the surface membrane and prevent an eruption.
5. Learn to remove eggs and egg dishes from the oven just before they are done. Let them stand several minutes, covered, to complete cooking.
6. Use an Amana Browning Skillet to fry eggs.
7. Yolks and whites cook at different rates. Scrambling minimizes the difference.
8. The addition of water slows and evens the cooking. Without water, one egg cooks in about 30 seconds to 1 minute.

Cheese Tips:

1. Add cheese to a dish at the end of the cooking time unless it is being used in a dish that is heated throughout.
2. Overcooking causes cheese dishes to become tough and rubbery.
3. Grated cheese makes a nice garnish for many dishes. Just sprinkle it over the cooked food, and return the dish to the Radarange Oven for just a few seconds.

Wake-up Special

Bouncy breakfast special for 6.

1 tablespoon butter or margarine
1/4 cup chopped green pepper

4 eggs
1/2 (10-3/4 oz.) can condensed cream of chicken soup

6 slices cooked and crumbled bacon
Salt
Pepper

1. Place butter in 2-quart casserole with green pepper. Sauté in Radarange Oven 30 seconds.
2. Stir eggs and soup together. Blend into sautéed mixture. Cook in Radarange Oven 4-1/2 to 5 minutes. Stir every 2 minutes.
3. Crumble bacon on top. Season with salt and pepper to taste.

MICRO-TIP: Mix remaining soup with 1/2 cup milk. Heat in Radarange Oven 1-1/2 to 2 minutes. Use as sauce for eggs.

Wake-up Special

Chinese Scramble

Bring an oriental dish for 4 to the breakfast table.

2 tablespoons butter or margarine

1. Heat butter in Radarange Oven in shallow 2-quart casserole 40 seconds.

6 eggs
1 (10 oz.) can drained Chinese vegetables
1/2 cup cooked rice
1/4 cup finely chopped onion
1/4 cup finely chopped green pepper
2 tablespoons soy sauce

2. Beat eggs. Combine eggs and remaining ingredients in casserole. Mix well.
3. Cook, covered, in Radarange Oven 4 minutes. Turn dish every minute.
4. Stir mixture from outside edge to center. Cook additional 1/2 to 1 minute. Let stand, covered, about 7 to 8 minutes before serving.

Eggs Delicious

A brunch dish for 4 that will never miss!

1/2 cup milk

1. Heat milk in Radarange Oven 45 seconds.

1 (3 oz.) pkg. cream cheese

2. Beat cream cheese until fluffy. Blend in milk.

1 large ripe tomato
6 medium eggs
1 cup diced, cooked ham
Salt
Pepper

3. Cut tomato in wedges. Stir in eggs, ham and tomato. Season to taste.

2 tablespoons butter or margarine

4. Melt butter in 10-inch pie plate 30 seconds. Pour egg mixture onto pie plate. Cook in Radarange Oven 7 to 8 minutes, stirring every 2 minutes.

MICRO-TIP: Cook 9 to 10 minutes for firmer eggs. Serve on or with hot buttered toast.

California Eggs

"A California dreamin'" of 2 servings.

4 eggs
1/2 cup fresh or canned mushrooms
1/4 teaspoon salt

1. Beat eggs until light and fluffy. Stir in mushrooms and salt.

1 peeled avocado
1 medium tomato
2 tablespoons butter

2. Cube avocado and cut tomato in wedges. Melt butter in 9-inch pie plate in Radarange Oven 30 seconds. Pour egg mixture into buttered dish. Cook in Radarange Oven 3-1/2 minutes. Stir every minute. Fold in avocados and tomato in last minute of cooking time.

Basque Eggs

A high hat breakfast for 4.

1/2 lb. diced, cooked ham
3 tablespoons chopped green pepper
3 tablespoons chopped onion
1 peeled and diced tomato
1/4 cup sliced pimiento-stuffed olives
1/8 teaspoon minced garlic
1 tablespoon butter

1. Sauté ham, green pepper, onion, tomato, olives and garlic in butter using 10-inch ceramic skillet. Cook in Radarange Oven 7 minutes. Stir every 2 minutes.

4 eggs
1/2 teaspoon salt
Dash pepper

2. Sprinkle salt and pepper on eggs. Beat mixture thoroughly. Place in skillet. Cook in Radarange Oven 4 minutes, or until eggs are set.

Scrambled Eggs Foo Yung Style

Eggs Foo Easy for 4.

2 tablespoons butter

1. Melt butter in 9-inch glass dish in Radarange Oven 30 seconds.

4 well-beaten eggs
3/4 cup well-drained bean sprouts
1/2 cup finely chopped onion
2 tablespoons chopped green pepper
1 teaspoon soy sauce
1/2 teaspoon salt
1/8 teaspoon pepper

2. Combine remaining ingredients. Place in glass dish. Cook in Radarange Oven 3-1/2 minutes. Stir lightly every minute.

Scrambled Eggs with Mushrooms

A morning mushroom mixture for 4.

2 cups chopped morel mushrooms

1. Place mushrooms in 10-inch ceramic skillet. Cook, covered, in Radarange Oven 5 minutes. Drain.

4 beaten eggs
2 tablespoons butter
1/2 teaspoon salt
Dash pepper

2. Stir in eggs, butter, salt and pepper. Cook in Radarange Oven 4 minutes. Stir every minute.

Cheese And Rice Fondue

Fondue is fun: 4 servings.

1 cup cooked rice

2 cups milk

4 well-beaten eggs
1 cup grated Cheddar cheese
1/2 teaspoon salt
1/4 teaspoon Worcestershire sauce
Few grains cayenne

1. Place rice in 2-quart casserole.
2. Pour milk over rice. Heat in Radarange Oven 3 minutes.
3. Stir in remaining ingredients. Cook in Radarange Oven 3 minutes. Stir halfway through cooking time.

MICRO-TIP: Serve this tasty fondue over toasted split corn muffins.

Savory Bacon Omelet

Sample this savory dish for 4.

2 tablespoons butter or margarine
1/2 cup thinly sliced scallions

8 beaten eggs
9-10 strips (1/2 lb.) cooked and crumbled bacon
1/4 cup heavy cream
2 teaspoons prepared mustard
1/4 teaspoon pepper
Salt to taste

1. Melt butter in 10-inch pie plate in Radarange Oven 30 seconds. Stir in scallions. Cook 1 minute.
2. Mix eggs, bacon, cream and seasonings. Pour into pie plate. Cook 4 to 6 minutes, stirring from outer edge to center every 2 minutes. Let stand 3 minutes. Fold. Turn onto warm platter.

Fried Egg

The All-American Breakfast.

1 egg
Butter or margarine

Method I: Break egg onto greased saucer or small plate. Pierce yolk with tines of fork. Cover with plastic wrap. Cook in Radarange Oven 30 to 60 seconds per egg, or to suit personal preference.

1 egg
1 tablespoon butter or margarine

Method II: Preheat 6-1/2-inch or 9-1/2-inch Amana Browning Skillet in Radarange Oven 1 minute. Melt butter in skillet. Break egg. Pierce yolk with tines of fork. Cook, covered, in Radarange Oven 30 to 60 seconds per egg. For repeated usage, preheat skillet 30 seconds.

Poached Egg

The popular poached egg.

1 cup water

1 egg

1. Boil water in Radarange Oven 1 to 1-1/2 minutes.
2. Break egg into 10-ounce custard cup. Pierce yolk with tines of fork. Pour water over egg. Cook, covered lightly with plastic wrap, 30 to 60 seconds per egg, or to suit personal preference. Remove with slotted spoon.

Ham And Cheese Bread Custard

A hearty lunch for 5.

6 slices firm white bread
1 (4-1/2 oz.) can deviled ham
6 slices process American cheese

1. Spread bread slices with deviled ham. Top with cheese slices. Arrange bread in 2-quart casserole, layering if desired.

1-1/2 cups hot milk
1 tablespoon minced onion
1 tablespoon Worcestershire sauce
1/2 teaspoon salt
1/8 teaspoon pepper
Dash cayenne
3 beaten eggs

2. Mix milk with onions and seasonings in 2-cup glass measure. Pour into egg mixture. Stir well. Pour over bread slices. Let stand 10 minutes. Cook, covered, in Radarange Oven 10 minutes. Turn dish quarter-turn every 2 minutes. Let stand, covered, 5 minutes before serving.

Deviled Egg Casserole

"Cute little devils" for 4.

4 hard-cooked eggs
2 tablespoons mayonnaise
2 tablespoons pickle relish
1/4 teaspoon salt
Dash pepper
Dash onion salt

1. Slice eggs in half, lengthwise and remove yolks. Mash yolks. Combine with mayonnaise, pickle relish, salt, pepper and onion salt. Spoon mixture into halves of egg whites.

1 (10 oz.) pkg. cooked broccoli

2. Spoon broccoli into greased 9-inch pie plate. Arrange eggs on top. Prepare cheese sauce.

2 tablespoons butter or margarine
2 tablespoons all-purpose flour
1/4 teaspoon salt
Dash pepper
1/4 teaspoon dry mustard

3. Melt butter in 1-quart casserole. Blend in flour, salt, pepper and mustard.

1 cup milk

4. Blend in milk gradually, stirring constantly. Cook in Radarange Oven 3 minutes. Stir every minute.

1/2 teaspoon Worcestershire sauce
1 (4 oz.) cup shredded natural Cheddar cheese

5. Stir in Worcestershire sauce and cheese. Pour cheese sauce over eggs and broccoli. Cook in Radarange Oven 5 minutes. Turn dish halfway through cooking time.

Crustless Quiche Lorraine

Fancy French brunch for 4.

9-10 (1/2 lb.) slices cooked and crumbled bacon
1 cup shredded Swiss cheese
1/4 cup minced onion

4 eggs
1 (13 oz.) can evaporated milk
3/4 teaspoon salt
1/4 teaspoon sugar
1/8 teaspoon cayenne pepper

1. Sprinkle bacon, cheese and onion in 9-inch glass pie plate.
2. Beat eggs, milk and seasonings with rotary beater until well blended. Pour over bacon mixture.
3. Bake in Radarange Oven 9 minutes. Stir every 3 minutes. Let stand 10 minutes before serving.

Swiss Cheese Fondue

4 servings of fine fondue.

4 slices white bread
Soft butter

1/2 lb. shredded Swiss cheese

2 eggs
1 cup milk
1/2 teaspoon dry mustard
1/4 teaspoon salt
1/4 teaspoon paprika
Dash cayenne

1. Spread slices of bread with butter. Arrange slices in 9 x 9 x 2-inch glass dish.
2. Sprinkle cheese over bread.
3. Beat eggs with remaining ingredients. Pour over bread and cheese. Cook in Radarange Oven 9 minutes until mixture is barely set. Turn dish every 2 minutes.

Onion Pie

6 servings: You'll "cry" for more.

9-inch baked pastry shell

2 tablespoons butter or margarine
3 cups thinly sliced sweet onions

1 lb. cottage cheese
1/3 cup whipping cream
2 egg yolks
1/2 teaspoon salt
1/8 teaspoon pepper

1. Prepare pie shell.
2. Melt butter in 9-1/2-inch Amana Browning Skillet in Radarange Oven 1 minute. Add onions. Cook in Radarange Oven 4 minutes. Stir every minute.
3. Warm cheese, cream and egg yolks in Radarange Oven 1 minute. Season, then pour into pie shell. Arrange onions over filling. Cook 10 minutes. Turn dish quarter-turn every 2 minutes. Let stand 5 minutes before serving.

Old Fashioned Farina Breakfast

Fashionable farina for 6.

4-1/2 cups boiling water
1/2 cup white farina
1 teaspoon salt

1. Mix farina and salt into 2-1/2 cups water in 2-1/2-quart casserole. Cook in Radarange Oven 30 seconds.

2 beaten eggs
1/4 cup sugar
1 tablespoon butter or margarine

2. Stir in eggs, sugar and butter. Mix until sugar dissolves.

1/4 cup golden seedless raisins
1/4 cup chopped toasted almonds
1 tablespoon toasted sesame seeds
1 teaspoon vanilla
1 teaspoon nutmeg

3. Stir in remaining ingredients. Blend in remaining 2 cups boiling water. Cook in Radarange Oven 8 to 9 minutes. Turn dish every 2 minutes.

Macaroni And Cheese

This delicious dish for 4 will put a feather in your cap.

1 (4 oz.) cup macaroni
1 quart hot water
1/2 teaspoon salt

1. Pour macaroni into 2-quart casserole with water and salt. Cook in Radarange Oven 10 minutes, or until macaroni is tender. Stir halfway through cooking time. Drain.

2 eggs
1 cup milk
1/2 teaspoon Worcestershire sauce
1/4 teaspoon salt
1/8 teaspoon paprika

2. Beat eggs slightly. Stir in milk and seasonings.

1-1/2 (6 oz.) cups grated Cheddar cheese

3. Layer macaroni and cheese in 1-1/2-quart casserole, ending with layer of cheese. Pour egg mixture over top. Sprinkle with additional paprika. Bake in Radarange Oven 4-1/2 minutes. Turn halfway through cooking time.

Hot Egg Scramble Mexicano

Best egg-vegetable combination north of the border: 6 servings.

2 tablespoons vegetable oil

1. Measure oil into 10-inch ceramic skillet. Heat in Radarange Oven 1 minute.

2 cups shredded iceberg lettuce
2 chopped ripe tomatoes
1 small, chopped, green pepper
1/4 cup chopped onion

2. Arrange lettuce, tomatoes, green pepper and onion in skillet.

6 beaten eggs
1 teaspoon chili powder
1 teaspoon salt
1/8 teaspoon pepper

3. Combine eggs with spices. Pour eggs over vegetables. Cook in Radarange Oven 8 to 10 minutes. Stir every 2 minutes. Let stand, covered, 5 minutes before serving.

Pretty Party Custards

Impress your table of 4 with beautifully baked custards, served in glamorous dessert dishes.

1-3/4 cup milk

3 eggs
1/4 cup sugar
1/8 teaspoon salt
1 teaspoon vanilla

Frozen or fresh strawberries, sliced peaches, raspberries or other fruit

1. Place 4 individual glass sherbet dishes, champagne glasses or any glass dishes of 6 to 8-ounce capacity into 9 x 9 x 2-inch glass baking dish.
2. Pour milk into glass measuring cup and heat in Radarange Oven until scalded, about 3-1/2 minutes.
3. While milk heats, beat eggs slightly and stir in sugar gradually. Blend well. Stir in salt, scalded milk and vanilla.
4. Pour custard into dessert dishes. Bake in Radarange Oven about 4-1/2 minutes, or until custards are barely set. Turn dishes every 30 seconds. Cool at room temperature. Chill, if desired. Prior to serving, top custards with partially defrosted frozen fruit or slightly sweetened fresh fruit.

Chocolate Baked Custard

8 colossal custards.

1/2 cup semi-sweet chocolate morsels
1-2/3 cups evaporated milk
1 cup water

4 eggs
1/2 cup sugar
1/2 teaspoon salt
1/2 teaspoon vanilla

1. Combine chocolate morsels, evaporated milk and water in 1-quart casserole. Cook in Radarange Oven 3 minutes, or until chocolate is melted.
2. Beat together eggs, sugar, salt and vanilla with rotary beater until well blended. Pour in chocolate mixture, stirring constantly. Beat until well blended. Pour into 8 custard cups. Set 4 cups in 9 x 9 x 2-inch glass dish. Pour 1 cup boiling water into dish.
3. Bake in Radarange Oven 5 minutes, or until custards are barely set. Turn dishes every 30 seconds. Repeat procedure with remaining custards.

Rice-Cheese Custard

Easy cheesy delight! 8 servings.

2 cups hot milk
1-1/2 cups shredded sharp Cheddar cheese
2 tablespoons butter or margarine
1 tablespoon minced onion
1/2 teaspoon salt
Dash paprika

2 beaten eggs
3 cups hot, cooked rice

1. Pour milk over cheese, butter, onion, salt, and paprika. Melt in Radarange Oven 3 minutes. Stir every minute.
2. Blend eggs and rice into cheese mixture. Stir well. Cook in Radarange Oven 8 minutes. Turn dish quarter-turn every 2 minutes. Cover. Let stand 10 minutes before serving.

Eggs and Cheese

Welsh Rarebit

Traditionally English! Rarebit serving 6.

1 cup beer or ale
1 tablespoon butter or margarine
1 lb. shredded sharp Cheddar cheese

1 beaten egg
1 teaspoon salt
1 teaspoon Worcestershire sauce
1/2 teaspoon dry mustard
1/2 teaspoon paprika
Dash cayenne

1. Pour beer into 2-quart casserole. Stir in butter, then cheese. Cook in Radarange Oven 1 minute.
2. Blend egg with seasonings. Stir in cheese mixture with wire whip. Cook 1 to 1-1/2 minutes in Radarange Oven.

MICRO-TIP: Stir well before serving over hot buttered toast.

Tomato Cheese Rarebit

6 servings of a "rare" taste treat.

1 (10-1/2 oz.) can tomato soup
1 lb. shredded sharp Cheddar cheese

2 separated eggs
1 teaspoon brown sugar
1/4 teaspoon paprika
Dash cayenne

1. Pour soup into 2-quart casserole. Blend in cheese. Cook in Radarange Oven 3-1/2 minutes. Stir halfway through cooking time.
2. Beat egg yolks, brown sugar, paprika and cayenne. Stir into cheese mixture. Cook in Radarange Oven 1 minute.
3. Whip egg whites until stiff. Fold into cooked mixture.

MICRO-TIP: Served best on hot toasted French bread.

Dried Chipped Beef Casserole

10 to 12 swell servings.

2 cups uncooked macaroni
2 (10-1/2 oz. each) cans cream of mushroom soup
2 cups milk
4 chopped hard-cooked eggs
1/4 cup chopped onion
1/2 lb. cubed longhorn cheese
1/4 lb. dried chipped beef

Combine all ingredients in 3-quart casserole. Refrigerate 12 to 15 hours. Cook, covered, in Radarange Oven 25 minutes. Stir every 10 minutes.

113
Soups and Sandwiches

Soup
Be creative with soups. Why not have a soup and sandwich party at which you allow your guests to select from a variety of convenience soup mixes. Remember, it will only take about 2 minutes to heat each cup of soup in the Radarange Oven. In this chapter you'll find recipes for light appetizer soups, as well as for the heartier "meal in a dish" soups. Recipes for chilled and cream soups are also included.

Sandwiches
Using hot sandwiches that teenagers can prepare themselves can be another way to "entertain with ease" with the Radarange Oven.

Frozen sandwiches
When reheating a frozen sandwich, defrost it in the Radarange Oven 1-2 minutes and let it stand briefly in order to allow for temperature equalization throughout the entire mass. Then heat for the necessary amount of time. The brief defrosting period before reheating will help to prevent an unevenness of reheating which may cause some areas to become overcooked and tough.

Micro-Tips for Sandwiches
* The filling determines the heating time.
* It's better to under-time, since overheating makes bread tough.
* Toasted bread and buns are less apt to become soggy than if untoasted.
* When using cheese, remember to add each slice during the last 30 seconds of heating time.

Iced Fresh Tomato Soup & Turkey "Leftover" in Buns

Manhattan Clam Chowder

4 lip-smackin' good servings.

2 strips bacon
1/2 cup finely chopped onion
1 cup finely chopped celery

3 tablespoons all-purpose flour

2 cups tomato juice
1 (8 oz.) bottle clam juice
1 (7-1/2 oz.) can minced clams
1/2 teaspoon powdered thyme
Salt
Freshly ground pepper

1. Place bacon in 3-quart glass casserole. Cook in Radarange Oven 2 minutes. Stir in onion and celery. Cook in Radarange Oven 2 minutes.
2. Blend in flour until smooth.
3. Mix in remaining ingredients. Cook in Radarange Oven 15 minutes. Stir every 5 minutes. Season to taste.

Oyster Chowder

4 simple servings to make — easy to enjoy.

1 (8 oz.) can oysters
Water

3 strips cooked and crumbled bacon
1/2 cup diced, cooked potato
1/4 cup diced, cooked onion
1 cup scalded milk
2 tablespoons butter or margarine
1 teaspoon salt
1/8 teaspoon nutmeg
1 bay leaf
1 teaspoon parsley

1. Drain oysters and reserve juice. Place oysters in 1-1/2-quart casserole. Add enough water to oyster juice to total 1-1/4 cups liquid. Cook in Radarange Oven 4 minutes, stirring every minute.
2. Mix in remaining ingredients. Cook in Radarange Oven 5 minutes. Remove bay leaf. Sprinkle with parsley.

Cheese Chowder

Chant for chowder: 6 servings.

1/2 cup finely chopped celery
1/2 cup finely chopped carrots
1/4 cup finely chopped onion
3 tablespoons butter or margarine

1/4 cup all-purpose flour
1/2 teaspoon salt
3 cups milk

2 cups chicken broth
1 cup grated sharp Cheddar cheese
Paprika

1. Mix celery, carrots and onions in butter in 2-quart casserole. Sauté in Radarange Oven 5 minutes. Stir halfway through cooking time.
2. Thoroughly mix in flour and salt. Gradually pour in milk, stirring constantly. Cook in Radarange Oven 5 minutes. Stir halfway through cooking time.
3. Stir in chicken broth and cheese. Cook 5 minutes. Stir halfway through cooking time. Sprinkle with paprika.

Lentil Soup

3 quarts of quick lentil soup.

1 cup dry lentils
8 cups water

1. Place lentils and 4 cups of water in 3-quart casserole. Bring to boil in Radarange Oven. Cook 2 additional minutes. Let stand, covered, 1 hour.

2 diced carrots
1 diced potato
1/2 cup minced onion
1/2 cup chopped celery
1 tablespoon salt

2. Pour 4 cups of water, carrots, potato, onion, celery, and salt into lentil casserole. Cook in Radarange Oven, covered, 15 minutes.

1 lb. pierced bratwurst

3. Stir casserole, then add sausage. Cook 20 minutes or until vegetables are barely tender. Remove and cool sausage.

2 tablespoons butter
2 tablespoons all-purpose flour

4. Melt butter. Blend in flour. Stir paste gradually into soup until smooth.
5. Skin, then slice sausages. Mix into soup mixture.

Fresh Zucchini Soup

4 to 6 zippy servings of zucchini soup.

3 cups (1-1/2 lbs.) zucchini

1. Trim stem and blossom ends of zucchini. Cut into 1-inch chunks. Place in 2-quart casserole.

1-1/3 cups water
2/3 cup condensed consomme
1/4 cup coarsely chopped onions
2 slices cooked and crumbled bacon
1 small clove garlic
2 tablespoons chopped parsley
1/2 teaspoon basil
1/2 teaspoon salt
1/2 teaspoon seasoning salt
1/8 teaspoon pepper

2. Gradually mix in remaining ingredients except cheese. Cook, covered, in Radarange Oven 15 minutes, or until zucchini is tender. Stir every 5 minutes. Cool slightly.

Grated Parmesan cheese

3. Blend zucchini mixture in blender, 2 cups at a time, until smooth. Return to casserole. Reheat to serving temperature if necessary. Sprinkle cheese over each serving.

Barley Soup

6 servings of barley soup for the sandwich crowd.

4 cups water

1. Bring water to boil in 2-quart casserole in Radarange Oven about 10 minutes.

4 beef bouillon cubes
1/2 cup quick barley

2. Melt bouillon cubes until dissolved. Mix in barley. Cook for 4 minutes. Stir halfway through cooking time. Let stand, covered, 10 minutes before serving.

Iced Fresh Tomato Soup

6 refreshing bowls of ripe-red soup.

2 tablespoons vegetable oil

1. Heat oil in 3-quart casserole in Radarange Oven 1-1/2 minutes.

1 cup chopped onions
6 chopped, medium ripe, tomatoes
1 (10-1/2 oz.) can condensed beef broth
1/4 cup catsup

2. Sauté onions and tomatoes 4 minutes. Stir in broth and catsup. Coo in Radarange Oven 2 minutes. Process in blender until smooth. Return mixture to casserole.

3 cups crushed ice
1 tablespoon dry dill weed
1 teaspoon salt
Dash Tabasco

3. Stir in ice and seasonings. Chill until cold. Blend with rotary beater before serving.

1/2 cup heavy cream

4. Whip cream until it stands in soft peaks. Use whipped cream to top each soup serving.

Chilled Pea Soup

A nice cool summer soup to serve 6.

2 (10 oz. each) pkgs. frozen green peas
1/2 cup finely chopped green onion
3/4 cup water
1/2 teaspoon salt
Dash nutmeg

1. Combine peas, onion, water, salt and nutmeg in 3-quart casserole. Cook, covered, in Radarange Oven 10 minutes. Stir halfwa through cooking time. Pour into blender. Process until smooth.

1 cup undiluted condensed chicken broth

2. Return mixture to casserole. Mix in broth. Cover and refrigerate.

1/2 cup heavy cream

3. Stir in cream just before serving.

Russian Borscht

4 servings of beet soup.

1/2 cup finely chopped onion
2 tablespoons butter

1. Sauté onion in butter for 3 minutes in Radarange Oven, using 3-quart casserole.

1 (16 oz.) can diced beets
2 cups hot water
3 bouillon cubes
1/2 teaspoon salt
Dash Tabasco

2. Process beets in blender. Mix in onions. Whir until smooth. Return vegetable pulp to casserole. Mix in hot water, bouillon cubes, salt and Tabasco. Heat in Radarange Oven 5 minutes.

1 tablespoon lemon juice
4 teaspoons sour cream

3. Stir in lemon juice. Separate into servings. Add sour cream to each serving.

Vegetable Beef Soup

3 quarts of "very vegetable" soup.

1 lb. ground beef
1/2 cup chopped onion

1. Crumble ground beef and onion in 4-quart ceramic Dutch oven. Cook, covered, in Radarange Oven 10 minutes. Stir halfway through cooking time.

8 cups water
1 tablespoon chopped fresh parsley
1 tablespoon salt
1/4 teaspoon pepper

2. Stir in water, parsley, salt and pepper. Cook in Radarange Oven 12 minutes. Stir.

2 cups canned tomatoes
1 cup medium egg noodles
1 cup diced potatoes
1 cup shredded cabbage
1/2 cup chopped celery
1/2 cup diced carrots
1/2 cup frozen peas or green beans

3. Mix in remaining ingredients. Cook 25 minutes, or until vegetables are barely tender. Stir halfway through cooking time.

Gourmet Endive Soup

A gourmet great! 4 to 6 super servings.

1/2 cup minced onion
2 tablespoons butter or margarine

1. Sauté onion in butter using 3-quart casserole. Cook in Radarange Oven 3 minutes.

4 firmly packed cups endive
2 (13-1/2 oz. each) cans chicken broth
Salt
Pepper

2. Trim and cut endive into onion mixture. Cook in Radarange Oven 8 minutes, until endive is tender. Stir in chicken broth. Season to taste.

MICRO-TIP: Top with Melba toast or French bread slices. Sprinkle with Parmesan cheese. If other greens are substituted for endive, cooking time may vary slightly. In this case, check tenderness of greens several times, adjusting cooking times accordingly.

Thick Hamburger Soup

3 quarts of a hamburger humdinger.

1 lb. ground beef

1. Brown meat in Radarange Oven 6 minutes in 4-quart ceramic Dutch oven. Stir every 2 minutes.

2 quarts water
1 (16 oz.) can stewed tomatoes
3 sliced onions
3 sliced carrots
3 stalks chopped celery
3 bouillon cubes
1/2 cup quick barley
1/4 cup pre-cooked rice
1 tablespoon salt
1 teaspoon MSG

2. Mix in remaining ingredients. Cook, covered, 45 minutes in Radarange Oven. Stir every 15 minutes. Let stand, covered, 30 minutes before serving.

Toasted Peanut Apple Sandwiches

4 breakfast sandwich specials.

1 (6 oz.) jar peanut butter
1/2 cup canned applesauce

8 slices frozen French toast

Soft butter or margarine

1. Blend peanut butter and applesauce.
2. Spread mixture on 4 slices toast. Top with remaining slices to make 4 sandwiches.
3. Butter sandwiches lightly on both sides.
4. Heat 9-1/2-inch Amana Browning Skillet in Radarange Oven 4-1/2 minutes. Place sandwiches in skillet.
5. Cook in Radarange Oven 2 minutes. Turn sandwiches over. Cook in Radarange Oven 3 minutes. Repeat with remaining 2 sandwiches.

   MICRO-TIP: May be eaten with fork. Syrup may be added if desired.

Reuben Sandwich

One colorful and flavorful taste treat.

Cooked corned beef
2 slices rye bread
2 tablespoons sauerkraut
1 slice Swiss cheese

1 teaspoon thousand island dressing

1. Thinly slice corned beef. Place corned beef on one slice of bread. Drain sauerkraut well. Place sauerkraut and Swiss cheese on corned beef.
2. Spread second piece of bread with dressing and place on top.
3. Heat sandwich on paper plate in Radarange Oven 45 seconds, or until cheese melts.

"Left-over" Turkey in Buns

4 better-than-ever "bunwiches".

4 toasted hamburger buns

1 cup chopped, cooked turkey
1 tablespoon instant minced onion
1 teaspoon instant minced parsley
1/4 cup chopped salted peanuts
1 hard-cooked chopped egg
1/2 cup mayonnaise
1/3 cup shredded sharp Cheddar cheese
Salt
Pepper

1. Place bottom halves of buns on paper towel-lined plate.
2. Combine remaining ingredients. Spread on buns.
3. Add bun tops to make 4 "bunwiches". Cook in Radarange Oven 3 minutes.

   MICRO-TIP: Any leftover cooked poultry or meat may be used.

Hot Swiss Chicken Salad Sandwiches

8 appetizing Alpine sandwiches.

1 cup diced cooked chicken
2/3 cup chopped celery
1/2 cup cubed natural Swiss cheese
1/4 teaspoon salt
1/4 cup mayonnaise
1/2 teaspoon lemon juice

8 hamburger buns

1. Combine all ingredients well, except buns.
2. At serving time, fill hamburger buns with 1/4 cup mixture. Heat in Radarange Oven. For one sandwich, heat 45 seconds. Heat 2 sandwiches 1 minute, 15 seconds. Heat 4 sandwiches 2-1/2 to 3 minutes.

Barbecued Mushroom Burgers

6 to 8 hearty servings.

1 lb. ground beef
1 cup chopped onion
1 cup diced celery

1. Combine beef, onion and celery in 1-1/2-quart casserole. Cook in Radarange Oven 7 minutes. Stir halfway through cooking time.

1 (10-3/4 oz.) can condensed onion soup
1 (10-3/4 oz.) can water
2 (4 oz. each) cans sliced, drained mushrooms
1/2 cup catsup
3 tablespoons quick-cooking tapioca
2 teaspoons chili powder

2. Add remaining ingredients. Stir well. Cook in Radarange Oven 8 minutes.

3. Let stand, covered, 10 minutes before serving.

MICRO-TIP: Serve on hamburger buns.

Creamed Dried Beef

A different dried beef dish for 2.

3 tablespoons butter

1. Melt butter in 1-quart glass casserole 1 minute in Radarange Oven.

3 tablespoons all-purpose flour
1-1/2 cups milk

2. Stir in flour. Blend to smooth paste. Cook in Radarange Oven 1 minute. Gradually stir in milk. Cook in Radarange Oven 4 minutes. Stir every 30 seconds.

1 (4 oz.) pkg. dried beef
Pepper

3. Tear dried beef into pieces. Stir into cooked sauce. Season with pepper to taste. Heat 1 minute in Radarange Oven before serving.

Circle Burgers

A-round for 4.

1 lb. ground beef
1 teaspoon salt

1. Combine meat and salt. Form into doughnut-shaped patties leaving holes in the centers. Place on Radarange Oven cooking grill and cook to degree desired, about 7 minutes. Turn patties over halfway through cooking time.

2 split English muffins
Prepared mustard
1/2 cup pasteurized process cheese spread

2. Toast muffins. Spread with mustard and top with patties. Spoon 2 tablespoons cheese into center of each patty.

3. Heat in Radarange Oven until cheese melts, about 1-1/2 minutes.

Toasted Crabmeat Sandwiches

Grab for crabmeat: 6 sandwiches.

1 cup flaked crabmeat
3 chopped, hard-cooked eggs
4 tablespoons mayonnaise
1 teaspoon minced onion
1 teaspoon Worcestershire sauce
1/4 teaspoon lemon juice
3/4 teaspoon salt
1/8 teaspoon pepper

1. Mix crabmeat, eggs, onion, Worcestershire sauce, mayonnaise, salt, pepper and lemon juice.

6 slices bread
1/4 cup shredded Cheddar cheese
Paprika

2. Toast bread. Spread mixture on toast. Sprinkle with cheese, then paprika. Heat in Radarange Oven 1 minute, or until cheese melts. Serve immediately.

Hot Swiss Tuna Salad Sandwich

8 salad sandwiches for a light supper.

1 (7 oz.) can drained, flaked white tuna
2/3 cup chopped celery
1/2 cup cubed natural Swiss cheese
1/4 teaspoon salt
1/4 cup mayonnaise
1/2 teaspoon lemon juice

1. Combine ingredients, blending well.

8 hamburger buns

2. Fill each bun with 1/4 cup tuna mixture. Heat each bun 45 seconds in Radarange Oven. Four sandwiches can be heated in 2-1/2 to 3 minutes.

One-of-A-Kind Soup

One-of-a-kind soup for an important occasion: 6 servings.

1 peeled, finely chopped potato
1/2 cup finely chopped onion
1 peeled, chopped cucumber
2 stalks chopped celery
1 peeled, chopped tart apple
1 tablespoon butter or margarine

1. Combine chopped vegetables, apple, and butter in 3-quart casserole. Cook in Radarange Oven 10 minutes. Stir halfway through cooking time.

4 cups hot chicken broth
1 cup light cream
1 teaspoon curry powder
1 teaspoon salt
Pinch white pepper
Chives

2. Stir in hot broth. Cook additional 10 minutes. Cool. Blend in electric blender until smooth. Stir in cream and seasonings. Chill. Serve cold. Sprinkle with chopped chives.

Hamburger-Stuffed French Bread

Hurry to bake this hamburger dish for 6.

1 (1 lb.) loaf French bread

1. Cut off ends of loaf. Hollow out center of bread leaving thick crust. Reserve 1-1/2 cups bread crumbs.

1-1/2 lbs. ground beef
3/4 cup chopped onion
3/4 teaspoon oregano
1/2 teaspoon salt
1/4 teaspoon pepper

2. Combine beef, onion, oregano, salt and pepper in 10-inch ceramic skillet. Cook, covered, in Radarange Oven 10 to 12 minutes. Stir halfway through cooking time. Drain.

1-1/2 cups French bread crumbs
1 slightly beaten egg
1-1/2 cups shredded Cheddar cheese
1/3 cup chopped parsley
1/4 teaspoon dry mustard

3. Mix in bread crumbs, egg, cheese, parsley and mustard. Blend thoroughly. Fill loaf with mixture. Replace cut ends and secure with toothpicks.

4. Divide loaf if necessary. Cover with plastic wrap. Cook in Radarange Oven 6 minutes. Turn halfway through cooking time.

King Size Sandwich

6 giant servings.

1 lb. ground beef
1/2 cup chopped onion

1. Combine meat and onion in 10-inch ceramic skillet. Cook in Radarange Oven 5 minutes. Stir halfway through cooking time.

1 (10-1/2 oz.) can vegetable soup
2 tablespoons catsup
1 teaspoon prepared mustard
1/2 teaspoon salt
1/4 teaspoon pepper

2. Stir in soup, catsup, and spices. Cook in Radarange Oven 5 minutes. Stir halfway through cooking time.

6 split toasted buns
6 slices tomato
6 slices onion

3. Spoon mixture onto buns. Top with tomato and onion slices.

3 halved slices American cheese

4. Cover sandwiches with cheese slices and bun tops. Heat in Radarange Oven 15 seconds, or until cheese melts.

Red Dog Sandwich

Red dog sandwiches for 6 robust eaters.

4 cubed skinless wieners
1/4 lb. cubed sharp Cheddar cheese
6 chopped stuffed olives
1 chopped hard-cooked egg
1 tablespoon minced onion
1 tablespoon chili sauce
1 tablespoon mayonnaise
6 hamburger buns

Combine ingredients except buns in large bowl. Mix well. Spoon 1/3 cup mixture into each bun. Heat in Radarange Oven 45 seconds, per sandwich.

MICRO-TIP: If sandwiches are prepared ahead of serving time, wrap them individually in plastic film. Refrigerate. Allow 60 seconds heating time per sandwich.

123
Vegetables

You'll be "cooking in color" when you prepare vegetables in the Radarange Oven. Vegetables require very little liquid to cook, thus they retain a truer vegetable color and texture. Vitamin retention is another plus when cooking in the Radarange Oven. This is due to the lesser amount of liquid used, and the shorter cooking time required.

Consider Ratatouille, Grecian Eggplant and Zucchini Casserole, which are just a few of the delightful recipes in this chapter. You will also find a timetable to guide you in defrosting frozen vegetables, and a timetable for cooking fresh vegetables.

For a change of pace, try basil with tomatoes, celery seed for beets and cabbage, and paprika as a garnish for cauliflower and corn. For more ideas to make your vegetable dishes extra special, refer to our spice and herb chart on the inside cover. Grated cheese is great on broccoli, asparagus and Brussels sprouts. Add cheese just before serving, and heat it in the Radarange Oven for 15 seconds, until melted.

Tips for cooking vegetables

Most vegetables will cook faster and more evenly if covered.

Be careful not to overcook vegetables. Vegetables cooked with microwaves will continue to cook after the Radarange Oven shuts off, so allow vegetables to stand, covered, for a few minutes before serving.

When cooking vegetables such as baked potatoes, arrange them in a circle and leave a space of about 1-inch between each. Turn potatoes during cooking.

To heat canned vegetables, drain some liquid from the can, empty the contents into a casserole, cover, and heat. (The excess liquid may be refrigerated for a few days and used in creating your own "Special" Soup.) Allow about 2 minutes of heating time for each cup of vegetable. Add seasonings after heating.

German Potato Salad

Heiszer Kartoffel Salat für 6.

4 cups sliced, cooked potatoes

6 slices bacon
1/2 cup chopped onion

2 tablespoons all-purpose flour
2 tablespoons sugar
1-1/2 teaspoons salt
1/2 cup vinegar

1 teaspoon celery seed
Parsley

1. Bake potatoes in Radarange Oven. Cool, peel and slice.
2. Cut bacon in small pieces. Cook in 10-inch ceramic skillet in Radarange Oven 5 minutes. Stir halfway through cooking time. Stir in onion. Cook 2 minutes.
3. Stir in flour, sugar and salt. Mix thoroughly. Slowly stir in vinegar. Cook in Radarange Oven 3 minutes. Stir every minute.
4. Blend in potatoes and celery seed. Toss lightly. Garnish with chopped parsley. Serve hot.

Scalloped Potatoes

5-6 scrumptious servings.

5-6 medium sliced potatoes

4-1/2 tablespoons all-purpose flour
1-1/4 teaspoons salt

1-1/2 cups scalded milk
3 tablespoons butter
Paprika

1. Arrange half of sliced potatoes in 8-inch glass baking dish.
2. Combine flour with salt. Sprinkle half on potatoes. Repeat steps 1 and 2 with remaining ingredients.
3. Pour milk over potatoes. Dot with butter. Sprinkle generously with paprika. Bake in Radarange Oven 20 minutes, or until potatoes are barely tender.

Potato Cups

4 servings of appealing potato cups.

4 small peeled potatoes
2 medium onions

1/2 cup melted butter or margarine
Salt
Pepper
Paprika

1. Slice potatoes and onions 1/8-inch thick. Alternate slices of potato and onion in four 6-ounce custard cups.
2. Spoon 2 tablespoons butter or margarine over each potato. Season to taste. Cook in Radarange Oven 3 minutes per potato. Sprinkle with paprika. Let stand 10 minutes before serving.

Party Potatoes

Potatoes for your party of 8.

8 to 10 potatoes
1/2 cup water

1 (8 oz.) pkg. cream cheese
1 (8 oz.) carton prepared French onion dip
1-1/2 teaspoons salt
1/8 teaspoon pepper
Garlic salt (optional)

Butter
Paprika

1. Peel potatoes. Cook in Radarange Oven with water in 4-quart casserole 20 minutes. Drain.
2. Beat together cream cheese and onion dip until well-blended. Stir in hot potatoes, one at a time, beating until light and fluffy. Stir in salt, pepper and garlic salt.
3. Spoon into 2-quart casserole. Dot with butter. Place in Radarange Oven 10 minutes, covered. Stir halfway through cooking time. Sprinkle with paprika.

Toll House Baked Beans

Toll House Baked Beans

6 to 8 servings of home cooked flavor and taste.

2 (1 lb. 2 oz. each) cans New England style baked beans
1/4 lb. cooked and crumbled bacon
1 (1 lb.) can solid pack tomatoes
1/2 cup minced onion
2 tablespoons dark molasses
1 tablespoon sugar
2 teaspoons dry mustard

1. Pour beans into 2-quart casserole or bean pot. Break up tomatoes and combine with remaining ingredients, mixing into beans.
2. Cover and cook in Radarange Oven 10 minutes. Stir halfway through cooking time. Let stand 10 minutes before serving.

Broccoli Special

6 to 8 servings of broccoli worth boasting about.

2 tablespoons minced onion
2 tablespoons butter or margarine

1-1/2 cups sour cream
2 tablespoons sugar
1 teaspoon vinegar
1/2 teaspoon poppy seed
1/2 teaspoon paprika
1/8 teaspoon salt
Dash pepper

2 (10 oz. each) pkgs. cooked broccoli
1/3 cup chopped cashews

1. In 1-quart casserole, sauté onion in butter in Radarange Oven 2 minutes. Stir.
2. Mix in sour cream, sugar, vinegar, poppy seed, paprika, salt and pepper. Heat in Radarange Oven 2 minutes. Stir halfway through cooking time.
3. Pour sauce over broccoli. Top with cashews.

Zucchini Casserole

Zesty zucchini for 6 to 8.

4 cups thinly sliced zucchini
1/4 cup hot water

3/4 cup shredded sharp cheese
1 egg
1/2 cup dairy sour cream
1 tablespoon all-purpose flour

3 slices cooked and crumbled bacon
1/4 cup browned bread crumbs

1. Place zucchini in 1-1/2-quart casserole with water. Cook, covered, in Radarange Oven 7 minutes. Drain.
2. Mix cheese, egg, sour cream and flour. Stir into zucchini. Cook in Radarange Oven 3 minutes. Stir.
3. Combine bacon and bread crumbs. Sprinkle over casserole. Cook in Radarange Oven 2 minutes.

Asparagus Sea Shore Style

A vegetable delicacy for 6.

2 (10 oz. each) pkgs. frozen asparagus pieces
1/4 cup water

1. Place asparagus and water in 2-quart glass casserole. Cook in Radarange Oven 10 minutes. Drain.

1 (10-3/4 oz.) can condensed cream of shrimp soup
1 (3 oz.) pkg. softened cream cheese
Dash cayenne pepper
1 (4-1/2 oz.) can small shrimp

2. Blend together soup, cream cheese and pepper. Drain shrimp and stir in. Pour over asparagus.

1/2 cup buttered crumbs
Paprika

3. Top with buttered crumbs. Sprinkle with paprika.
4. Cook in Radarange Oven 8 minutes. Turn dish halfway through cooking time.

Creamy Green Beans And Mushrooms

5-6 servings good green beans.

1 lb. fresh green beans
1/4 cup water

1. Rinse and drain beans. Trim ends. Snap in two. Place in 1-1/2-quart glass casserole with water. Cook in Radarange Oven 8 minutes. Let stand 2 minutes. Drain.

1 (3 to 4 oz.) can drained, sliced mushrooms
1 cup dairy sour cream
1 tablespoon brown sugar
1 tablespoon all-purpose flour
1/2 teaspoon salt
Dash pepper

2. Stir in mushrooms. Mix remaining ingredients. Stir into beans and mushrooms. Cover.
3. Cook in Radarange Oven 3 to 4 minutes. Stir. Let stand, covered, 5 minutes before serving.

Green Bean Treat

4-6 delectable vegetable servings.

2 pkgs. frozen green beans
1/2 teaspoon salt
1 can condensed celery soup

1. Place beans in 1-1/2-quart covered casserole. Sprinkle with salt. Spread soup over beans.

1 can French fried onion rings

2. Bake in Radarange Oven 5 minutes. Break up beans with fork and stir. Place onion rings on top. Continue to cook in Radarange Oven for 8 minutes.

Honeyed Beets

6 servings sure to please the "honey" in your life.

1 tablespoon cornstarch
1/2 teaspoon salt
1 tablespoon water or beet juice
2 tablespoons vinegar
1/4 cup honey
1 tablespoon butter or margarine

1. Combine cornstarch and salt in small glass bowl or measure. Blend in water or beet juice. Stir in vinegar, honey and butter.
2. Cook in Radarange Oven 1 minute 15 seconds, or until thickened. Stir every 20 seconds.

2 cups diced or sliced cooked beets

3. Place beets in 1-1/2-quart covered casserole. Pour sauce over beets. Heat 1-1/2 minutes in Radarange Oven. Let stand about 10 minutes to blend flavors. Heat 1 more minute.

Snappy Baked Beans

6 snappy servings.

4 slices bacon

1. Cut bacon in small pieces. Cook in large 1-1/2-quart casserole or bean pot in Radarange Oven 4 minutes. Stir halfway through cookir time.

2 (1 lb. each) cans pork and beans
1/4 cup dark corn syrup
1/4 cup catsup
2 tablespoons chopped onion
3/4 cup finely crushed ginger snaps

2. Stir in remaining ingredients. Bake in bean pot in Radarange Oven 10 minutes. Stir every 4 minutes.

Calico Bean Pot

12 huge servings to fill the heartiest of appetites.

8 slices bacon

1. Cut bacon in small pieces. Cook in Radarange Oven until brown and crisp. Remove meat and reserve.

1 cup chopped onion

2. Sauté onion in bacon fat 3 minutes. Stir halfway through cooking time.

1 (1 lb.) can drained green beans
1 (1 lb.) can drained lima beans
1 (1 lb. 15 oz.) can drained pork and beans
1 (1 lb.) can drained kidney beans
3/4 cup firmly packed brown sugar
1/2 cup vinegar
1/2 teaspoon garlic salt
1/2 teaspoon dry mustard
1/8 teaspoon pepper

3. Combine all ingredients. Mix lightly. Pour into 3-quart casserole. Bak in Radarange Oven 20 minutes. Stir and turn dish every 7 minutes. Mixture may be divided between two 1-1/2-quart casseroles and frozen for later use. Cooking time will be cut to 13 to 15 minutes fo each 1-1/2 quart casserole.

Ratatou

Mandarin Carrots

6 colorful servings to brighten your dinner table.

4 cups carrots
2 tablespoons butter

1. Cut carrots in 2-inch strips. Place carrots and butter in 1-1/2-quart casserole. Cook in Radarange Oven 10 to 12 minutes. Turn dish and stir halfway through cooking time.

1 (11 oz.) can drained mandarin orange sections
1/2 teaspoon salt
1/8 teaspoon ginger

2. Add mandarin oranges, salt and ginger. Cook in Radarange Oven 2 minutes.

3. Let stand 2 or 3 minutes before serving.

Broccoli And Carrots Au Gratin

Broccoli and carrots for company.

1 (10 oz.) pkg. frozen chopped broccoli
1 (10 oz.) pkg. frozen sliced carrots in butter sauce

1. Thaw vegetables in Radarange Oven just to separate. Place vegetables in 1-1/2-quart casserole.

1/2 cup non-dairy creamer
3 tablespoons all-purpose flour
1/4 cup shredded Parmesan cheese
1/2 teaspoon salt
1/8 teaspoon pepper
1 cup boiling water or chicken broth

2. Combine non-dairy creamer, flour, cheese, salt and pepper. Sprinkle over vegetables. Cover with boiling water or broth.

3. Place lid on casserole and cook in Radarange Oven 8 minutes. Stir halfway through cooking time. Stir, cover and let stand 5 minutes.

Sweet Sour Red Cabbage

6 "rosy-red" servings.

1 shredded medium red cabbage
2 peeled, chopped, tart apples
1 cup boiling water
1/2 cup apple cider vinegar
3 tablespoons butter
3 tablespoons sugar
1/2 teaspoon salt
1 stick cinnamon

Combine ingredients in 2-1/2-quart covered casserole. Cook in Radarange Oven 10 to 12 minutes until cabbage is barely tender. Stir every 3 to 4 minutes.

Ratatouille

(Rä tä tyū́ ē)

Fancy French vegetable dish
4 to 6 servings.

1-1/2 cups peeled, diced eggplant
1/2 cup thinly sliced onion
1 clove minced garlic
3 tablespoons olive oil

1. Place eggplant, onion, garlic and oil in 2-quart casserole. Cook in Radarange Oven 5 minutes.

1 medium green pepper
1-1/2 cups sliced zucchini

2. Cut green pepper into 1/2-inch strips. Layer peppers and zucchini over eggplant mixture.

1 (16 oz.) can stewed tomatoes
1 teaspoon salt
1/4 teaspoon Italian seasoning
Dash pepper

3. Add seasonings to tomatoes. Pour over vegetables. Cook in Radarange Oven 8 to 10 minutes. Let stand 5 minutes before serving.

Cauliflower Oriental

6 Eastern servings.

1 medium head cauliflower

1. Wash cauliflower and remove outer green stalks. Place in 2-quart covered casserole. Cook in Radarange Oven 10 minutes. Drain.

1/2 cup chopped onion
1/2 cup diced celery
3 sprigs chopped parsley
1 tablespoon butter or margarine

2. In 1-1/2-pint ceramic dish, sauté onion, celery, and parsley in butter. Cook in Radarange Oven 5 minutes. Stir halfway through cooking time.

1 beef or chicken bouillon cube
1 cup hot water
1 tablespoon cornstarch
1 tablespoon soy sauce
Dash pepper

3. Dissolve bouillon cube in water. Blend in cornstarch, soy sauce and pepper. Pour into onion mixture. Cook in Radarange Oven 2 minutes. Stir every 30 seconds. Place cooked cauliflower head on serving dish. Top with sauce.

Boston Style Baked Corn

6 servings favored-in-the-Commonwealth.

1 cup catsup
2 tablespoons brown sugar
1 teaspoon dry mustard
1/2 teaspoon salt
1/4 cup chopped onion
2 (12 oz. each) cans drained whole kernel corn

1. Combine catsup, brown sugar, mustard and salt in 1-1/2-quart casserole. Stir in onion and corn. Cook in Radarange Oven 5 minutes. Stir halfway through cooking time.

3 slices cooked and crumbled bacon

2. Sprinkle bacon pieces over top. Cook in Radarange Oven 5 minutes.

Peas And Lettuce Medley

4-5 melodious servings of peas.

2 cups fresh peas
1/4 small head, coarsely chopped, iceberg lettuce
1/4 cup sliced green scallions
1/2 teaspoon sugar

1. Place peas, lettuce, onions, and sugar in 1-1/2-quart casserole. Coo[k] in Radarange Oven 5-1/2 minutes. Stir halfway through cooking time.

3/4 teaspoon salt

2. Stir in salt. Let stand, covered, 5 minutes before serving.

Baked Peas

6 servings of peas from the pods.

2 (10 oz. each) pkgs. frozen peas

1. Place peas in 1-1/2-quart casserole. Cook, covered, in Radarange Oven 6 minutes. Drain.

6 slices cooked and crumbled bacon
1 cup light cream
1/2 teaspoon salt
1/8 teaspoon pepper
Pinch celery seed
1/2 cup buttered, browned bread crumbs

2. Stir in bacon, cream and seasonings. Sprinkle with crumbs.

3. Cook in Radarange Oven 6 minutes.

Baked Broccoli Fondue

Fondue for you to do: 6 servings.

1-1/3 cups hot milk
1-1/3 cups soft white bread crumbs
1/4 cup butter or margarine
1/2 teaspoon salt

1. Mix crumbs, butter and salt into milk.

4 separated eggs

2. Beat egg yolks. Stir into milk mixture. Transfer to 2-quart casserole. Cook in Radarange Oven 45 seconds. Stir every 15 seconds.

1-1/2 cups well-drained, fresh, chopped broccoli
2/3 cup shredded sharp Cheddar cheese

3. Mix in broccoli and cheese. Cool slightly. Beat egg whites until stiff. Fold into mixture. Cook in Radarange Oven 9 minutes. Turn dish quarter-turn every 3 minutes. Let stand 3 to 4 minutes before serving.

Baked Potatoes

Radarange Oven potatoes are best ever.

1 to 4 (7 oz. each) baking potatoes

1. Select uniform, medium-size baking potatoes, about 7 oz. each. Scrub potatoes well.
2. Pierce each potato all the way through with large fork. Arrange potatoes on paper towel. Leave about 1-inch space between potatoes. Avoid placing 1 potato in center, surrounded by other potatoes.
3. Bake in Radarange Oven. Bake 1 potato 4 to 6 minutes. Bake 2 potatoes 8 to 11 minutes. Bake 4 potatoes 16 to 19 minutes. The exact time varies according to size and variety of potatoes. Turn potatoes halfway through cooking time.

   MICRO-TIP: If potatoes feel slightly firm after recommended cooking time, allow a few minutes standing time. Potatoes will finish cooking on their own.

Baked Potato Boats

4 potato pleasers.

4 (7 oz. each) potatoes

1. Bake potatoes as directed above. After removing from Radarange Oven, let stand a few minutes.
2. Cut thin slice through skin from top of each potato. Scoop out potato from skins, leaving thin, unbroken shell from each potato.

2 tablespoons butter
1 cup milk
Salt
Pepper

3. Mash potatoes. Stir in butter, milk, salt and pepper. Whip until potato mixture is light and fluffy.

1/2 cup shredded sharp cheese
Paprika

4. Lightly spoon potato mixture back into shells. Top each potato with cheese. Sprinkle with paprika.
5. Place potatoes in a circle on paper plate. Cook in Radarange Oven 5 minutes.

   MICRO-TIP: If desired, stuffed potatoes may be prepared ahead and refrigerated. Heat in Radarange Oven 6 to 7 minutes.

Lemon Potatoes

4 generous helpings to please the potato eaters in your home.

1-1/2 lbs. (3 large) peeled, quartered potatoes

1. Arrange potatoes in 8 x 8 x 2-inch buttered baking dish.

2 teaspoons lemon juice
2 teaspoons melted butter

2. Combine lemon juice with butter and spread over potatoes.

3 tablespoons shredded Parmesan cheese
2 teaspoons fresh grated lemon peel
1/2 teaspoon paprika

3. Combine cheese, lemon peel and paprika. Sprinkle over potatoes.

4. Cover and cook in Radarange Oven 12 to 14 minutes. Turn dish every 3 to 4 minutes.

Sweet Potatoes in Orange Shells

6 sweet servings of a colorful favorite.

3 large oranges
Orange juice

1. Cut oranges in half. Juice oranges. Measure 3/4 cup juice. Set aside. Remove membrane from orange halves, wash shells and reserve.

4 medium sweet potatoes

2. Scrub potatoes. Prick with fork. Place on paper towel. Bake 15 to 18 minutes in Radarange Oven until potatoes are well done. Turn potatoes halfway through cooking time.

1 tablespoon butter
1/4 cup firmly packed brown sugar
1/2 teaspoon salt
3/4 cup orange juice

3. Peel hot potatoes. Mash and season with butter, brown sugar, and salt. Whip until light and fluffy. Gradually blend in orange juice.

6 large marshmallows

4. Place sweet potato mixture in orange shells. Set in 1-1/2-quart glass casserole. Top each serving with marshmallow. Bake in Radarange Oven 5 minutes.

Candied Sweet Potatoes

4 sensational servings of sweet potatoes.

3 or 4 large sweet potatoes

1. Wash sweet potatoes. Pierce each potato with fork. Place on paper towel in Radarange Oven 8 minutes, or until almost tender. Peel potatoes. Slice in half-inch slices.

1/4 cup butter
1/4 cup firmly packed brown sugar
3-4 tablespoons prepared mustard

2. Melt butter in 10-inch ceramic skillet. Blend in sugar and mustard. Stir in potatoes.

3. Cook in Radarange Oven 5 minutes. Turn potato slices over halfway through cooking time.

Acorn Squash Marmalade

2 acorn halves.

1 acorn squash

1. Pierce skin of squash with sharp knife in many places. Bake in Radarange Oven 4 minutes. Cut squash in half lengthwise. Remove seeds.

3 tablespoons margarine
3 tablespoons orange marmalade
Cinnamon

2. Combine softened margarine and marmalade. Blend well. Spoon mixture into center of squash. Sprinkle with cinnamon.

3. Continue baking in Radarange Oven 3 minutes, until tender.

Cranberry Squash

Cranberry creation for two.

1 acorn squash

1. Cut squash in half, lengthwise. Remove seeds.

1/4 cup fresh cranberries
1/4 cup firmly packed brown sugar
1 tablespoon butter or margarine
1/4 teaspoon cinnamon

2. Combine cranberries, brown sugar, butter and cinnamon. Fill squash halves with this mixture.

3. Place squash in shallow glass baking dish. Bake in Radarange Oven 10 minutes. Turn dish halfway through cooking time. Test with fork.

Squash Souffle

4-6 servings of squash.

1 (12 oz.) pkg. frozen squash
1 egg
1/2 cup milk
2 tablespoons melted fat
1 tablespoon finely chopped onion
1 cup bread crumbs
1 teaspoon salt

1. Thaw squash. Mix all ingredients together except cheese and 1/2-cup bread crumbs. Place in buttered 1-quart souffle dish or casserole.

1/2 cup shredded cheese

2. Sprinkle with cheese and remaining bread crumbs.

3. Bake in Radarange Oven 7 minutes. Turn dish every 2 minutes.

Hot Chili-Cheese Tomatoes

20 tremendous tomato halves.

10 medium, peeled tomatoes

1. Cut tomatoes in half. Squeeze each gently to drain off some excess liquid and seeds.
2. Arrange 5 tomatoes (10 halves) cut side up in 2-quart shallow bakin dish.

1 cup dairy sour cream
1/2 teaspoon salt
1/4 teaspoon pepper
1 teaspoon sugar
2 tablespoons chopped green onion
1 tablespoon all-purpose flour
2 tablespoons chopped canned green chilies
1 cup shredded Longhorn or Monterey Jack cheese

3. Thoroughly blend sour cream with seasonings, onion, flour and chopped chilies. Spoon 1 teaspoon mixture over each tomato. Sprinkle 1/2 cup cheese over all.
4. Cook in Radarange Oven 4 minutes. Turn dish halfway through cooking time. Repeat procedure with remaining tomatoes.

Stewed Tomatoes

4 to 6 servings of tasty tomatoes.

2 cups boiling water
3 medium tomatoes

1. Pour boiling water over tomatoes to loosen skins. Drain. Peel tomatoes. Cut in fourths.

1 tablespoon minced onion
2 teaspoons sugar
1 teaspoon salt
1/8 teaspoon pepper
2 tablespoons butter or margarine

2. Combine tomatoes, onion, seasonings and butter in 1-1/2-quart glass casserole. Cook 3 minutes in Radarange Oven.

1/2 cup bread cubes

3. Stir in bread cubes. Cook 2 to 3 minutes in Radarange Oven.

Sweet-Sour Red Cabba

Tomato Casserole

A tremendous tomato treat for 6.

4 tablespoons melted butter
2 tablespoons chopped onion

2 cups canned tomatoes
1/2 cup cracker crumbs
1 beaten egg
1/2 cup cubed Cheddar cheese
1 tablespoon sugar
1 teaspoon salt
1/4 teaspoon paprika
Dash chili powder

1. Sauté onions in butter in 10-inch ceramic skillet. Cook in Radarange Oven 1-1/2 minutes.
2. Combine tomatoes, crumbs, egg, cheese, sugar, salt and spices with onion mixture. Transfer to greased 1-1/2-quart casserole. Cook in Radarange Oven 5 minutes. Stir every 1-1/2 minutes.

Creamed Radishes

Ravishing radishes for 4.

2 cups radishes
1/4 cup water

1 tablespoon butter or margarine
1 tablespoon all-purpose flour
1/4 teaspoon curry powder
3/4 cup milk
Salt

1. Halve radishes. Place in 1-quart casserole. Cook, covered, with water in Radarange Oven 8 minutes. Stir halfway through cooking time.
2. Melt butter in 2-cup glass measure. Stir in flour and curry powder. Pour milk into flour mixture. Cook in Radarange Oven 3 minutes. Stir every minute, or until mixture thickens. Pour over radishes. Season to taste.

Scalloped Rutabaga And Apple Casserole

Rush to serve rutabaga and apples for 5 to 6.

1-1/4 lbs. peeled, diced rutabaga
1/4 cup water

1 tablespoon butter
Dash salt

1/2 cup sliced, pared apple
1/2 cup firmly packed brown sugar
Pinch of cinnamon

1/3 cup all-purpose flour
2 tablespoons butter

1. Cook rutabaga in 2-quart casserole with water for 15 minutes in Radarange Oven.
2. Mash butter with salt. Stir into rutabaga. Spread half of mixture into 1-1/2-quart casserole.
3. Mix apples, 1/4 cup brown sugar and cinnamon. Sprinkle over rutabaga. Top with remaining rutabaga.
4. Mix flour, remaining brown sugar and butter until crumbly. Sprinkle on top. Bake in Radarange Oven 10 minutes. Turn dish halfway through cooking time.

Spicy Carrots

Spice up your table with 4 servings of colorful carrots.

4 cups carrots
2 tablespoons butter or margarine

2 tablespoons brown sugar
1 teaspoon dry mustard
2 drops Tabasco sauce
Salt
Pepper

1. Cut carrots in 2-inch strips. Place in 1-1/2-quart casserole with butter. Cook in Radarange Oven 5 minutes. Stir halfway through cooking time.
2. Combine brown sugar, mustard and Tabasco sauce. Pour over carrots. Cook in Radarange Oven 5 minutes, or until tender. Stir halfway through cooking time. Season to taste.

Sweet-Sour Vegetable Relish

Delicious hot or cold: 4 sweet-sour servings.

1 beaten egg
1/2 cup sour cream
1/4 cup vinegar
2 tablespoons sugar
1/2 teaspoon salt
1/4 teaspoon pepper
Dash of cayenne

1. Combine egg, sour cream, vinegar, sugar and spices in 2-cup glass measure. Cook in Radarange Oven 3 minutes. Stir thoroughly every minute. Cool, and then refrigerate.

1 (10 oz.) pkg. frozen mixed vegetables
1/4 cup water

2. Place vegetables in small covered casserole with water. Cook in Radarange Oven 7 minutes. Stir halfway through cooking time. Drain. Cool.

3/4 cup chopped celery

3. Stir in celery and 1/3 cup of liquid mixture. Toss lightly.

MICRO-TIP: Serve salad on crisp lettuce leaf. Garnish with paprika. Salad dressing yields 2/3 cup. Refrigerate remainder. Or prepare a double recipe of salad using 2 (10 oz. each) packages of vegetables and 1-1/2 cups of celery.

Fried Mushrooms

4 fabulous servings of morels.

1/4 lb. butter or margarine

1. Preheat 9-1/2-inch Amana Browning Skillet 2-1/2 minutes in Radarange Oven. Melt butter 1 minute.

12 large morel mushrooms
2 well-beaten eggs
1 cup cracker meal

2. Dip mushrooms in eggs, and then cracker meal. Drain. Fry until golden brown in Radarange Oven 2 to 3 minutes. Turn halfway through cooking time.

MICRO-TIP: Fried mushrooms may be frozen, then reheated when needed on double thickness of paper towel.

French Onion Casserole

Viva la onion casserole: 4-6 servings.

4 medium, sliced onions
3 tablespoons butter or margarine

1. Sauté onions in butter using 1-1/2-quart casserole in Radarange Oven 7 minutes. Turn dish halfway through cooking time.

2 tablespoons all-purpose flour
Dash pepper
3/4 cup beef bouillon
1/4 cup dry sherry

2. Blend in flour and pepper. Pour in bouillon and sherry. Cook until thickened. Stir every 30 seconds.

1-1/2 cups plain croutons
2 tablespoons melted butter
1/2 cup shredded process Swiss cheese
3 tablespoons grated Parmesan cheese
Paprika

3. Toss croutons with butter. Spoon onto onion mixture. Sprinkle with cheeses, and then paprika. Cook in Radarange Oven 3 minutes.

Fresh Vegetable Timetable

Add 1 to 2 tablespoons water to most fresh vegetables and cook in covered casserole in Radarange Oven. Cream or butter may also be used for moisture.

FRESH VEGETABLES	COOK	STIR
Asparagus (1 lb.)	6 minutes	Every 2 minutes
Beans, Green (3 cups) (Pole Beans)	10-12 min.	Every 3 min.
Beans, Green (1 lb.)	8 minutes	At 4 minutes
Beans, Yellow Wax (1 lb.)	13 minutes	Every 3 minutes
Beets (1-1/2 lbs.)	15 minutes	
Broccoli (1-1/2 lbs.)	11 minutes	At 6 minutes
Brussels Sprouts (1-1/2 lbs.)	8 minutes	At 4 minutes
Cabbage, red or white (1-3/4 lbs. med. head)	11 minutes	Occasionally
Cabbage, Chinese celery (1 lb.)	8 minutes	
Carrots (1 lb. whole)	8 minutes	
Cauliflower (1 lb. whole) (1 lb. flowerets)	10 minutes 8 minutes	
Celery (4 cups)	7-9 minutes	
Corn on the Cob (4 ears)	8 minutes	
Eggplant (4 cups, diced)	5 minutes	
Mushrooms (1/2 lb.)	4 minutes	At 2 minutes
Okra (1 lb.)	5 minutes	
Parsnips (4 medium)	8-10 minutes	
Peas, green (2 cups, 2 lbs.) (4 cups, 4 lbs.)	6 minutes 9-10 minutes	At 3 minutes At 5 minutes
Pea Pods (Chinese) (1 lb.)	10 minutes	Every 3 minutes
Potatoes, Red boiled (4 large, 2 lbs.)	12-15 minutes	At 6 minutes
Potatoes, New boiled in Jackets (6,2-inch in diameter, 1/4 cup water)	10 minutes	At 5 minutes
Potatoes, Sweet (2 med.) (4 med.)	8-11 minutes 15-18 minutes	
Rutabaga (3 cups, cubed)	10 minutes	Every 3 minutes
Spinach (1/2 lb.)	4 minutes	
Squash, Acorn (1—1-1/2 lbs.)	8 minutes	At 4 minutes, split, remove seeds, cook 4 minutes.
Squash, Butternut (1 lb.)	8 minutes	
Squash, Summer Yellow Crookneck, (5 small, about 1 lb.)	6-1/2 minutes	
Swiss Chard (3/4 lb., 1/4 cup water)	7 minutes	
Tomatoes, Baked (1 lb., 2 medium)	3-4 minutes	
Turnips (1 bunch)	10 minutes	At 5 minutes
Zucchini (2 med., 3 cups)	7 minutes	At 4 minutes

Colcannon

Scotch-Irish cabbage and potato dish to delight 4.

1 cup chopped onion
1/4 cup butter or margarine

2 cups seasoned mashed potato
1 cup chopped, cooked cabbage

Salt
Pepper
1/2 cup toasted bread crumbs
1/2 cup shredded sharp cheese

1. Place onion and 2 tablespoons butter in shallow baking dish. Cook in Radarange Oven 4 minutes. Stir every minute.
2. Stir in potato and cabbage. Mix well. Cook, covered, in Radarange Oven 1-1/2 minutes.
3. Sprinkle with salt, pepper and crumbs. Dot with remaining butter. Top with cheese. Cook in Radarange Oven 2 minutes, or until cheese melts.

Frozen Vegetable Timetable

Times given are for one 10-ounce package. For two 10-ounce packages, cooking time will be almost doubled.

FROZEN VEGETABLE	RADARANGE TIME
Asparagus, green spears	5—6 minutes
Broccoli	8-1/2—9-1/2 minutes
Beans, Green Cut Beans, Wax	7-1/2—8-1/2 minutes
Beans, Green French Cut	7—8 minutes
Beans, Fordhook Lima	8-1/2—9-1/2 minutes
Cauliflower	5—6 minutes

FROZEN VEGETABLE	RADARANGE TIME
Corn, Cut	4—5 minutes
Corn on Cob	5—6 minutes
Mixed Vegetables	6-1/2—7-1/2 minutes
Okra	6—7 minutes
Peas, Green	4-1/2—5-1/2 minutes
Peas & Carrots	5—6 minutes
Spinach	4-1/2—5-1/2 minutes

Okra

Simple okra to serve 4.

1 lb. fresh okra

1. Wash and sort okra. Leave small pods whole. Cut large pods into 1-inch pieces.

2 tablespoons water
2 tablespoons butter or margarine
Salt
Pepper

2. Heat water and butter in 1-1/2-quart covered casserole in Radarange Oven 1 minute. Add okra. Cook, covered, in Radarange Oven 5 minutes, or until tender. Season to taste with salt and pepper.

New Potatoes and Peas in Cream

4 servings of new potatoes from a new recipe!

1 (10 oz.) pkg. frozen peas
1 lb. small new potatoes

1. Defrost peas. Scrub potatoes. Cut off band of skin around middle of each. Pierce each potato twice with fork. Put potatoes in medium-size cooking bag. Fold open end of bag under potatoes loosely. Cook in Radarange Oven 8 minutes. Turn bag over halfway through cooking time.

2 tablespoons butter or margarine
1 tablespoon all-purpose flour
1/2 teaspoon salt
1/8 teaspoon pepper
1/2 cup light cream

2. Melt butter in 2-1/2 quart casserole. Stir in flour, salt and pepper. Blend in light cream gradually, stirring constantly. Cook in Radarange Oven 1 minute, 30 seconds. Stir every 30 seconds.

1/3 cup chopped onion

3. Stir peas and onions into sauce. Mix well. Add cooked potatoes. Stir to coat all. Cook in Radarange Oven 3 minutes.

143
Breads, Cereals & Pasta

Start the day right with cereal prepared in the Radarange Oven. 1/3 cup quick-cooking oats requires only the addition of 2/3 cup water, a dash of salt, and can be cooked in the serving dish for 1 minute, 15 seconds.

Breads and rolls heat nicely in the Radarange Oven. Remember to wrap them in paper towels or napkins to prevent soggy crusts.

Rice can be cooked with a minimum of effort. Use regular, long grain or white rice. Precooked rice requires only the addition of boiling water. Allow it to stand, covered, several minutes.

Most rice dishes require only 1/4 the time of conventional cooking. For your first pleasant experience with rice, try this Instant Rice Pudding.

Instant Rice Pudding

3 cups milk
1 cup instant rice
1/4 teaspoon salt
1 (3-1/4 oz.) pkg. vanilla pudding

Mix milk, rice, salt, vanilla pudding in 8 x 8 x 2-inch glass dish. Cook, covered, in Radarange Oven 10 minutes. Stir 3 times. Let stand for 10 minutes.

We're sure you will enjoy the pasta recipes in this chapter.

Timings for pasta convenience food items such as packaged lasagna or macaroni often differ due to many variables. To cook these foods from the frozen state, simply add a short amount of time to the cooking period.

Always be sure to remove any food items packaged in metal trays or foils from their containers. Place them in utensils suitable for use in the Radarange Oven.

Allowing foods to stand just a few minutes after the cooking time is advisable. These few minutes permit the moisture molecules in the foods to equalize.

If you choose to cook your foods on the Slo Cook or Automatic Defrost Cycle, remember to increase your total cooking time 2-2-1/2 times, depending on the model of your Radarange Oven.

Hot Mexican Rice

Honey Bran Kuchen

9 to 12 pieces of kuchen from your kitchen.

3/4 cup all-purpose flour
2-1/2 teaspoons baking powder
1/2 teaspoon cinnamon
1/2 teaspoon nutmeg
1/4 teaspoon salt

1. Sift together flour, baking powder and spices.

2 cups 40% bran flakes

2. Stir in 1-1/2 cups bran flakes.

1/2 cup milk
1 well-beaten egg
1/4 cup honey
3 tablespoons melted shortening

3. Combine milk, egg, honey, and shortening. Pour in flour mixture. Moisten.

1/4 cup firmly packed brown sugar
2 tablespoons melted butter or margarine

4. Mix brown sugar, butter and remaining bran flakes.

5. Pour batter into 8 x 8 x 2-inch baking dish. Top with sugar mixture. Bake in Radarange Oven 4-1/2 minutes. Turn dish quarter-turn every 1-1/2 minutes.

Perpetual Muffins

6 dozen "perpetually good" muffins.

4 cups all-bran cereal
2 cups 100% bran cereal
2 cups boiling water

1. Pour hot water over cereals.

1 quart buttermilk
3 cups sugar
4 beaten eggs
1 cup soft shortening or oil
5 cups all-purpose flour
5 teaspoons baking soda
1 teaspoon salt

2. Stir in buttermilk, sugar, eggs, shortening, flour, soda and salt in that order. Stir until just blended. Refrigerate.

3. Remove from refrigerator. Without stirring batter, fill paper baking cups half full. Arrange 6 cups in circle. Bake in Radarange Oven 2-1/2 minutes. Remove muffins immediately from custard cups to cooling rack.

MICRO-TIP: Mixture can be stored in the refrigerator for 3-4 weeks.

Carolina Corn Bread

Corn bread from the land of cotton.

1 cup yellow corn meal
1 cup sifted all-purpose flour
1/4 cup sugar
4 teaspoons baking powder
1/2 teaspoon salt

1. Sift together corn meal, flour, sugar, baking powder and salt.

1 egg
1 cup milk
1/4 cup soft shortening
Vegetable spray-on coating

2. Stir in egg, milk and shortening. Beat with rotary beater until smooth. Spray inside of 9-1/2-inch Amana Browning Skillet with vegetable spray-on coating. Preheat skillet 2 minutes. Pour batter into skillet. Bake in Radarange Oven 5-1/2 to 6 minutes. Turn halfway through cooking time. Cool 10 minutes. Turn onto platter.

Date And Nut Bread

Make a date to make this loaf!

1 cup chopped dates
3/4 cup chopped pecans
1-1/2 teaspoons baking soda
1/2 teaspoon salt

1. Grease 8-1/2 x 4-1/2 x 2-1/2-inch glass loaf dish. Combine dates, pecans, baking soda and salt.

3/4 cup boiling water
3 tablespoons shortening

2. Mix water and shortening in lightly. Let stand 1 hour.

2 eggs
1/2 teaspoon lemon extract
1/2 teaspoon vanilla
1 cup sugar
1-1/2 cups sifted all-purpose flour

3. Beat eggs lightly. Stir in remaining ingredients. Blend in date mixture with fork. Mix with rotary beater until well-blended. Pour into loaf dish. Cook in Radarange Oven 8 minutes. Turn dish quarter-turn every 2 minutes. Let stand in dish 10 minutes before removing to cooling rack.

MICRO-TIP: Even more delicious when refrigerated overnight before slicing.

Fancy Corn Bread

"Fancy-dancy" corn bread.

1 cup yellow corn meal
1 cup all-purpose flour
1/4 cup sugar
4 teaspoons baking powder
1/2 teaspoon salt
1/4 teaspoon chili powder

1. Sift corn meal, flour, sugar, baking powder, salt and chili powder.

2 beaten eggs
1 cup milk
1/4 cup oil
1 (12 oz.) can drained Mexicorn niblets

2. Stir in eggs, milk and oil. Beat with rotary beater until smooth. Mix in corn. Pour batter into 9-inch round dish. Bake in Radarange Oven 5-1/2 to 6 minutes, or until toothpick inserted in center comes out clean.

Radarange Microwave Muffin Bread

A toaster treat: 2 loaves.

4-1/2 to 5 cups all-purpose flour
2 (1/4 oz. each) pkgs. active dry yeast
1 tablespoon sugar
2 teaspoons salt

1. Combine 3 cups flour, yeast, sugar and salt in large mixing bowl.

2 cups milk
1/2 cup water

2. Warm milk and water in Radarange Oven 1 to 1-1/2 minutes. Stir into flour mixture. Beat with mixer until smooth. Stir in enough remaining flour to make stiff batter.

3. Cover bowl. Place in warm place. Let batter rise until light and doubled in bulk, about 1 to 1-1/2 hours.

2-3 tablespoons melted butter

4. Stir down yeast batter. Divide batter between two greased 8-1/2 x 4-1/2 x 2-1/2-inch loaf dishes. Cover, and let rise until doubled in bulk, about 1 to 1-1/4 hours. Cook in Radarange Oven 6-1/2 minute per loaf. Turn quarter-turn every 2 minutes. Brush top with butter. Cool 5 minutes. Remove from loaf dishes. Butter sides and bottom of loaves.

MICRO-TIP: To serve, slice, then toast in toaster.

Biscuit Breakfast Ring

Ring to sing about: 4 to 5 servings.

1/3 cup firmly packed brown sugar
3 tablespoons butter or margarine
1 tablespoon water
1/3 cup chopped nuts

1. Combine sugar, butter and water in 1-1/2-quart round baking dish or ceramic ring mold. Cook in Radarange Oven 1 minute. Stir until butter melts. Stir in nuts.

10 refrigerator biscuits

2. Cut each biscuit into fourths. Stir into sugar mixture, coating each piece. If using round dish, set custard cup or glass in center.

3. Cook in Radarange Oven 3 to 3-1/2 minutes. Turn dish quarter-turn every 30 seconds. Remove from Radarange Oven. Let stand 2 minutes. Invert on platter. Serve immediately.

Rolls

Tips and timings for rolls.

Do you find yourself wasting a lot of bread or pastry items in your home? You need never throw away day-old bread or dessert items again.

The Radarange Oven refreshes and restores the original flavor and texture of foods. Just be careful not to dehydrate by overheating.

1 sweet roll, dinner roll or prepared bun

1. Wrap roll in paper towel. Place in Radarange Oven 15 to 20 seconds.

2 sweet rolls, dinner rolls or prepared buns

1. Wrap rolls in paper towel. Place in Radarange Oven 25 to 30 seconds.

6 sweet rolls, dinner rolls or prepared buns

1. Wrap rolls in paper towel or place in straw basket lined with paper towels. Heat in Radarange Oven 1 to 1-1/2 minutes.

efrigerator Biscuits

Quick for quotable comments: 1 to 10 biscuits.

1 roll of 10 refrigerator biscuits

Method I: Preheat 9-1/2-inch Amana Browning Skillet in Radarange Oven 2-1/2 minutes. Bake 5 biscuits at one time. Bake in Radarange Oven 1-1/2 minutes. Turn biscuits over halfway through baking time. Preheat skillet 1-1/4 minutes. Repeat baking using same process.

1 roll of 10 refrigerator biscuits
1 tablespoon butter

Method II: Preheat 9-1/2-inch Amana Browning Skillet in Radarange Oven 2-1/2 minutes. Melt butter in skillet. Arrange 5 biscuits in skillet. Cook, covered, in Radarange Oven 2 minutes. Turn biscuits over halfway through cooking time.

Pancakes & Waffles

Batter beauties.

Do you need to fix pancakes and/or waffles on a moment's notice? No need to hesitate when you know you can use your Radarange Oven.

Next time you prepare pancakes or waffles conventionally, make a double batch. Freeze the extras. Place a double sheet of freezer or waxed paper between the pieces for easy separation without defrosting.

Pop them into the Radarange Oven whenever you desire. 4 frozen, pre-cooked pancakes or waffles require about 1-1/2 minutes heating time.

Don't forget to warm your pancake syrup in the Radarange Oven. Warmed syrup adds that "extra special" little touch to your breakfast table.

Pancakes, Complete Mix

7 to 8 perfect pancakes.

1 cup pancake mix
3/4 cup water or milk

1 tablespoon vegetable oil

1. Stir water into mix. Preheat 9-1/2-inch Amana Browning Skillet in Radarange Oven 2-1/2 minutes.
2. Line skillet with oil. Pour batter on skillet to form 4 pancakes. Cook in Radarange Oven 1 minute. Turn pancakes over halfway through cooking time.

MICRO-TIP: For repeated usage of skillet, preheat 1-1/2 minutes before pouring in additional batter.

Frozen Waffles

Waffles that taste wonderful!

1 frozen waffle

Preheat 9-1/2-inch Amana Browning Skillet in Radarange Oven 2-1/2 minutes. Place waffle on skillet. Cook in Radarange Oven 2 minutes per waffle. Turn waffle halfway through cooking time.

MICRO-TIP: For repeated usage of skillet preheat 1-1/4 minutes each time in Radarange Oven.

Blueberry Coffee Cake From Muffin Mix

8 servings to make you smile.

1 (13 oz.) pkg. blueberry muffin mix

1. Prepare muffin mix according to package directions. Pour mixture into 9-inch round cake dish. Cook in Radarange Oven 6 minutes. Turn dish halfway through cooking time.

Soft butter or margarine
2 tablespoons brown sugar
1/2 teaspoon cinnamon

2. Brush warm coffee cake with butter. Sprinkle sugar and cinnamon over top.

Gingerbread Coffee Cake

Serve this cake for 8 without reservation.

1 cup all-purpose flour
1/2 cup sugar
1/2 teaspoon baking soda
1/4 teaspoon salt
1/4 teaspoon cinnamon
1/4 teaspoon ground ginger

1. Sift flour, sugar, soda and spices into mixing bowl.

1/4 cup soft shortening

2. Cut in shortening until mixture is crumbly. Reserve 1/2 cup of the crumbs for topping.

1 beaten egg
1/2 cup cultured buttermilk
1 tablespoon dark molasses

3. To remaining mixture in bowl, stir in egg, buttermilk and molasses. Mix until blended. Spread batter in 9-inch round dish. Bake in Radarange Oven 5 minutes. Sprinkle reserved crumb mixture on top. Turn halfway through cooking time.

Butter Pecan Coffee Cake

Pick the pecan coffee cake to enhance your table of 8.

1 (14 oz.) pkg. Butter Pecan Coffee Cake Mix
1/3 cup butter
2 tablespoons milk

1. Blend packaged topping mix, butter and milk. Bring to boil in Radarange Oven for 2-1/2 minutes.
2. Follow package directions for mixing cake mix.
3. Pour batter into 9-inch round cake dish. Bake in Radarange Oven 6 minutes. Turn dish every 2 minutes. Spread topping mix over cake. Serve warm.

Cinnamon Rolls

A "sin" to eat, but quite a treat.

2 cups warm water
1/3 cup sugar
2 pkgs. active dry yeast

1. Place water in 3 or 4-quart bowl. Sprinkle sugar and yeast on top. Stir until dissolved.

1/3 cup liquid shortening
2 beaten eggs
1 tablespoon salt

2. Blend in shortening, eggs and salt.

6-6-1/2 cups all-purpose flour
Soft butter

3. Beat in flour, 2 cups at a time. Mix well after each addition. Scrape dough from sides of bowl to center. Cover, and let stand 20 minutes. Turn out on floured board. Divide dough. Roll into 1/4-inch thick rectangle. Spread with butter.

1/2 cup firmly packed brown sugar
2 tablespoons cinnamon

4. Sprinkle with cinnamon and sugar. Roll dough tightly. Cut in 3/4-inch slices. Place in greased dish for about one hour. Bake in Radarange Oven 5 minutes. Turn quarter-turn every minute. Let stand 8 to 10 minutes.

MICRO-TIP: Pecan rolls may be prepared as above, except place 3/4-inch slices into dish with 1/4 cup each of butter, brown sugar and pecan pieces. Melt butter in round glass cake dish. Stir in brown sugar. Sprinkle with pecans. Place rolls in dish and bake as above.

Cheese Batter Bread

2 loaves of better batter bread.

4-3/4 to 5-1/2 cups sifted all-purpose flour
3 tablespoons sugar
4 teaspoons salt
2 pkgs. active dry yeast

1. Mix 1-3/4 cups flour, sugar, salt and yeast in large bowl.

1 cup milk
1 cup water
2 tablespoons butter or margarine

2. Heat milk, water and butter in 1-quart measure in Radarange Oven 2 minutes. Gradually stir liquid into dry ingredients. Beat 2 minutes at medium speed. Scrape sides of bowl occasionally.

1-1/2 cups shredded sharp Cheddar cheese
1 egg

3. Thicken batter with cheese, egg and 3/4 cup flour. Beat at high speed 2 minutes. Gradually stir in enough flour to make stiff batter that leaves sides of bowl. Cover with damp cloth. Let rise until double in bulk, about 50 to 60 minutes. Stir batter vigorously 1/2 minute. Turn into two deep, greased 1-quart casseroles. Bake each in Radarange Oven 8 minutes. Turn every 2 minutes. Cool 10 minutes.

Basic Long Grain Rice

6 servings of rice that's nice.

2-1/2 cups boiling water
1 cup rice
1 teaspoon salt

Pour water over salt and rice in 2-quart casserole. Cook, covered, in Radarange Oven 7 minutes. Let stand, covered, 10 minutes. Fluff with fork before serving or adding to casserole dishes.

Fluffy Orange Rice

Colorful and tasty; a pleasure to serve 4 guests.

1 cup chopped celery
1/4 cup chopped onion
1/4 cup butter

1. Sauté celery and onion in butter using 9-1/2-inch Amana Browning Skillet in Radarange Oven. Cook 5 minutes in Radarange Oven.

1-1/4 cups water
2 tablespoons orange juice concentrate
1/2 teaspoon salt
1-1/3 cups packaged, pre-cooked rice

2. Stir in water, concentrate and salt. Boil. Mix in rice. Cover, and let stand 5 minutes.

Parsleyed Rice

Be partial to parsleyed rice: 4 to 6 servings.

2 cups cooked rice
1/2 lb. shredded American cheese
1 tablespoon instant minced onion
1 chopped clove garlic
1/2 teaspoon salt
1/8 teaspoon pepper

1. Combine rice, cheese, onion and spices.

1/4 cup olive oil
2 eggs

2. Beat olive oil with eggs. Stir into rice mixture.

2 cups milk
1 tablespoon parsley flakes

3. Stir in milk. Bake in Radarange Oven 15 minutes. Turn dish quarter-turn every 5 minutes. Sprinkle with parsley flakes.

Spanish Rice

spicy Spanish servings.

4 to 6 slices bacon

1. Cook bacon. Reserve 2 tablespoons drippings.

1 (16 oz.) can tomatoes
3/4 cup water
1 (6 oz.) pkg. Spanish rice mix

2. Combine tomatoes, water, bacon drippings and Spanish rice mix. Cook, covered, in Radarange Oven 8 minutes. Stir halfway through cooking time. Allow to stand, covered, 5 minutes. Crumble bacon over top before serving.

Pre-cooked Rice

Add some spice to your life with rice.

Equal measures of pre-cooked rice and water
1/2 teaspoon butter per serving
1/8 teaspoon salt per serving

Bring water, butter and salt to boil in Radarange Oven 1-1/2 to 9 minutes depending on quantity of water required. Stir in rice. Cover. Let stand 5 minutes. Fluff with fork to serve.

MICRO-TIP: For a firmer rice, use 1 tablespoon less water per serving.

Fancy Cornbread

Hot Mexican Dish

Unforgettable fiesta for 8.

1 cup chopped onion
1 medium diced green pepper
1 small chopped hot chili pepper
1 minced clove garlic
1/2 cup raw long grain rice
1/4 cup vegetable oil

1. Preheat 9-1/2-inch Amana Browning Skillet 4-1/2 minutes in Radarange Oven. Sauté onion, green pepper, chili pepper, garlic and rice in oil in Radarange Oven 3 minutes. Stir every minute.

1 lb. ground beef

2. Mix in ground beef. Cook in Radarange Oven 3 minutes. Stir every minute.

1 (16 oz.) can stewed tomatoes
3/4 cup seedless raisins
1/4 cup pine nuts
1 tablespoon chili powder
2 teaspoons salt
1/4 teaspoon seasoned pepper

3. Stir in remaining ingredients. Cook, covered, in Radarange Oven 9 to 10 minutes. Let stand, covered, 10 minutes. Fluff with fork before serving.

Quick Fried Rice

Good aroma, good tasting, and good eating: 4-5 servings.

2 tablespoons chopped onion
2 tablespoons butter or margarine

1. Sauté onion in butter in 9-1/2-inch Amana Browning Skillet in Radarange Oven.

2 cups quick rice
1-2/3 cups water
2 beef bouillon cubes
1/2 teaspoon salt
Dash pepper

2. Stir in rice, water, bouillon cubes, salt and pepper. Cook, covered, in Radarange Oven 7 minutes. Stir well halfway through cooking time. Stir and let stand, covered, 5 minutes, until water is absorbed.

2 eggs
1/4 cup chopped green onion
2 teaspoons soy sauce

3. Mix in slightly beaten eggs and green onion. Cook 1-1/2 minutes. Stir every 30 seconds or until eggs are set. Stir in soy sauce.

Polenta

6 "plenty popular" portions of polenta!

3 cups corn meal mush
1 cup shredded Cheddar cheese
1/8 teaspoon paprika
Dash cayenne
All-purpose flour

Mix cheese, paprika and cayenne with corn mush. Cook in Radarange Oven 1 minute. Let stand, covered, 5 minutes. Pour into greased 8 x 8 x 2-inch dish. Cover and refrigerate. At serving time, coat each piece with flour. Sauté in hot bacon drippings or oil.

Green Rice Casserole

8 great portions.

3 cups cooked long grain rice
2 cups firmly packed, finely chopped fresh spinach or 1 (10 oz.) pkg. frozen spinach
3/4 cup finely chopped green onion with tops
1/2 cup minced parsley
1/2 cup slivered almonds
1/4 cup butter or margarine
2 teaspoons lemon juice
1 teaspoon salt
Dash garlic powder

1. Combine rice, spinach, onion, parsley, almonds, butter, lemon juice, salt and garlic powder in 2-quart casserole.

1 egg
1 cup milk

2. Beat egg with milk. Stir into rice mixture. Cover and refrigerate. When thoroughly cool, bake in Radarange Oven 15 minutes. Turn dish every 4 minutes. Let casserole set 10 minutes before serving.

Nibbles Snack

Don't quibble; serve nibbles! A great cereal snack for parties!

3/4 cup butter
2 teaspoons garlic salt
2 teaspoons onion salt
2 teaspoons celery salt
3 tablespoons Worcestershire sauce

1. Melt butter in large bowl. Add salts and sauce. Mix well.

1 small box thin pretzel sticks
1 (6-3/4 oz.) can cocktail peanuts
1 can mixed nuts or cashews
2 cups toasted wheat cereal squares
2 cups toasted rice cereal squares
2 cups round toasted oat cereal

2. Stir in remaining ingredients. Toss until well-coated.

3. Place in Radarange Oven 8 minutes. Mix thoroughly every 2 minutes. Cool. Store in plastic or tin container with tight-fitting lid.

155
Breads, Cereals, Pastas

Lasagna

12 servings of luscious lasagna.

1/2 lb. lasagna noodles

1. Cook noodles, following directions in recipe at bottom of this page, except increase cooking time to 10 minutes in Radarange Oven.

1 lb. ground beef
1 minced clove garlic

2. Cook ground beef and garlic in 10-inch ceramic skillet in Radarange Oven 6 minutes. Stir every 2 minutes.

1 envelope dry onion soup mix
1-1/2 cups water
1 (6 oz.) can tomato paste
1 (8 oz.) can tomato sauce
1/2 teaspoon salt
1/2 teaspoon sugar
1/4 teaspoon pepper
1 teaspoon oregano

3. Blend in onion soup, water, tomato paste, tomato sauce, salt, sugar, pepper and oregano. Stir well. Cook, covered, in Radarange Oven 10 minutes. Stir every 3 minutes.

1/2 lb. mozzarella cheese
1 lb. cottage cheese
Parmesan cheese

4. Layer meat mixture, noodles, mozzarella cheese and cottage cheese in 2-quart glass utility dish. Repeat layers, ending with meat sauce. Sprinkle top generously with Parmesan cheese. Bake in Radarange Oven 20 minutes. Turn dish every 4 minutes.

Noodles Romanoff

Oodles of noodles for 5.

4 cups water

1. Boil water in Radarange Oven 9 minutes using 4-cup measure.

1 (6-1/4 oz.) pkg. Noodles Romanoff
1 teaspoon salt

2. Place "boil-in-bag" Noodles Romanoff in 2-quart casserole with salt. Pour boiling water over pouch. Cook in Radarange Oven 6 minutes. Let stand, covered, 10 minutes. Drain. Remove noodles from pouch. Place in casserole.

2 tablespoons margarine
1/3 cup milk

3. Mix in margarine, Noodles Romanoff sauce mix, milk and packaged seasonings. Return to Radarange Oven 2 minutes. Stir after 1 minute. Serve immediately.

Macaroni Supreme

4 supreme servings!

1/4 cup chopped onion
1 tablespoon margarine

1. Sauté onion in margarine, using 1-quart casserole. Cook in Radarange Oven 2 minutes.

1 (10-1/2 oz.) can cream of celery soup
1 cup shredded Cheddar cheese
1/2 cup milk

2. Stir in soup, 3/4 cup cheese, and milk. Cook in Radarange Oven 3 minutes.

2 cups cooked macaroni
2 tablespoons buttered bread crumbs

3. Mix in macaroni. Top with remaining cheese and crumbs. Heat in Radarange Oven 6 minutes. Turn dish every 2 minutes.

Macaroni, Spaghetti, Noodles

Pasta for 6 people.

4 cups water
2 cups macaroni, spaghetti or noodles

1. Bring water to boil in Radarange Oven. Place pasta in 2-1/2-quart casserole.

1 tablespoon salt

2. Measure salt, and pour boiling water into casserole. Cook in Radarange Oven 6 minutes. Let stand, covered, 10 minutes. Drain. Rinse with warm water before serving.

asagne

Radarange Regular Oatmeal

4-6 simple servings.

1-1/2 cups quick-cooking rolled oats

1. Place oatmeal in 2-quart casserole.

3 cups room temperature water
3/4 teaspoon salt

2. Stir in water and salt. Cook in Radarange Oven 8 minutes. Stir halfway through cooking time. Let stand 4 to 5 minutes before serving.

Single Serving Oatmeal

Just the right breakfast for cold chilly mornings.

Method I:

1/4 cup quick-cooking rolled oats

1. Pour oatmeal into serving bowl.

1/2 cup room temperature water
1/8 teaspoon salt

2. Stir in water and salt. Cook in Radarange Oven 1 minute, 15 seconds. Stir, cover, and let stand few minutes before serving.

Method II:

3/4 cup water
1/8 teaspoon salt

1. Place water and salt in ceramic bowl or glass cereal bowl. Bring to boil in Radarange Oven.

1/3 cup quick-cooking rolled oats

2. Stir oats into boiling water. Return to Radarange Oven for 15 seconds, being careful not to allow water to boil over. Stir. Cook in Radarange Oven an additional 15 seconds.

MICRO-TIP: Cereal may be refrigerated overnight and quickly reheated in Radarange Oven for a speedy, hot breakfast.

Cornmeal Mush

6 to 8 servings of Southern style.

1 cup white or yellow cornmeal

1. Place cornmeal in 2-quart casserole.

4 cups room temperature water
1-1/2 teaspoons salt

2. Stir in water and salt. Cook in Radarange Oven 10 minutes. Stir half way through cooking time. Cover, and let stand 5 to 6 minutes before serving.

Easy Rice Pudding

Fluffy and flavorful treat for 4.

1 (2-3/4 oz.) pkg. egg custard mix
2 cups milk

1. Pour egg custard mix into 1-quart glass casserole. Gradually stir in milk until mixture is smooth.

1-1/2 cups cooked rice
1/2 cup raisins

2. Stir in rice and raisins. Mix well.

Nutmeg

3. Cook, uncovered, in Radarange Oven about 4 minutes. Stir well. Cook 3 additional minutes. Stir again. Sprinkle nutmeg over top of pudding. Cool or chill until set.

Peach Conserve, Strawberry Preserves and Grape Jel

Sauces, Jams & Jellies

Sauces
You will find sauces a real joy to make in your Radarange Oven. The double-boiler, which results in steam-burned hands, is gone forever. We've included some basic French sauces, as well as dessert and meat sauces. They can be made very easily in glass measuring cups, and require only minimum amounts of stirring. We suggest doing egg and cream sauces on the Slo Cook Cycle. For ease in stirring, most sauces are cooked uncovered.

Jams and Jellies
Jams and jellies are a delightful addition to any table! The Radarange Oven simplifies forever the usual long, drawn-out process of stirring and mixing. Surprise your family with flavorful jellies and jams you can make only in your Radarange Oven. Original-gift-givers love these recipes. They are such unique, yet always appreciated presents.

Quick Hollandaise

1/2 cup festive Hollandaise Sauce.

1/3 cup butter

1. Place butter in small casserole. Heat in Radarange Oven 45 seconds.

2 tablespoons lemon juice
2 egg yolks
1/4 teaspoon salt

2. Stir in lemon juice and egg yolks. Beat with whisk until well-mixed. Cook in Radarange Oven 60 seconds. Whisk thoroughly every 15 seconds. Stir in salt halfway through cooking time.

Hot Fudge Sauce

1 cup of sauce for sundaes!

1/2 cup sugar
3 tablespoons cocoa
1-1/2 tablespoons cornstarch
Dash salt

1. Mix dry ingredients in 1-quart casserole or 2-cup measure.

1/2 cup water
2 tablespoons butter or margarine

2. Stir in water, using water of room temperature. Cook in Radarange Oven 1-1/2 minutes. Stir every 30 seconds. Blend in butter or margarine. Cook in Radarange Oven for additional 30 seconds. Stir halfway through cooking time.

1 teaspoon vanilla

3. Remove from Radarange Oven. Stir thoroughly. Blend in vanilla.

White Sauce

1 cup of creamy white sauce.

1 tablespoon butter or margarine
1 tablespoon all-purpose flour
1/2 teaspoon salt

1 cup milk

1. Melt butter in 1-quart bowl or casserole in Radarange Oven 30 seconds. Stir in flour and salt to make smooth paste.
2. Blend in milk gradually, stirring constantly. Cook in Radarange Oven 3 minutes. Stir well at 30-second intervals.

MICRO-TIP: The above recipe makes thin sauce. For medium sauce, add 1 additional tablespoon each of butter and flour. For thick sauce, add 3 additional tablespoons each of butter and flour.

Bechamel Sauce

1 cup of seasoned sauce.

2 tablespoons butter or margarine

2 tablespoons all-purpose flour
1/2 teaspoon salt
1/2 cup chicken stock
1/4 cup cream

1 teaspoon grated onion
1/8 teaspoon white pepper
Dash of thyme

1. Melt butter in 1-quart glass casserole in Radarange Oven 30 seconds.
2. Stir in flour and salt. Blend to smooth paste. Mix chicken stock and cream. Stir in gradually.
3. Cook in Radarange Oven 1 minute. Stir well. Cook 1-1/2 to 2 minutes longer, stirring every 30 seconds.
4. Season with onion, pepper and thyme.

Hollandaise Sauce

1 cup of the all time—any time popular sauce.

1/4 cup butter or margarine

1/4 cup light cream
2 egg yolks
1 tablespoon lemon juice
1/2 teaspoon dry mustard
1/4 teaspoon salt
Dash Tabasco

1. Melt butter in 4-cup glass measure in Radarange Oven 1 minute.
2. Stir in remaining ingredients. Cook in Radarange Oven 1 minute. Stir every 15 seconds.
3. Stir briskly with wire whip until light and fluffy.

Sour Cream Dressing

2-1/2 cups dressing to "dress" your salad.

1 egg
1 cup milk
1/2 cup sugar
3 tablespoons all-purpose flour
1 teaspoon dry mustard

1. Combine first 5 ingredients. Cook in Radarange Oven 3 minutes or until mixture thickens. Stir after 2-1/2 minutes.

1/2 cup vinegar
1 tablespoon butter or margarine

2. Stir in vinegar and butter. Cook in Radarange Oven 1 minute. Stir halfway through cooking time. Cool.

1 teaspoon salt
1 cup dairy sour cream

3. When mixture is cold, stir in salt. Fold in sour cream.

MICRO-TIP: May be served over vegetable salad or lettuce wedges.

Brown Gravy From Meat Drippings

1 cup of bargain brown gravy.

2 tablespoons grease separated from drippings
1-1/2 tablespoons all-purpose flour

1. Mix grease and flour in 2-cup glass measure. Stir until smooth. Coo in Radarange Oven 2 to 3 minutes or until browned.

1 cup drippings
1/4 teaspoon MSG
Salt
Pepper

2. Stir in remaining ingredients. Return to Radarange Oven 1 to 2 minutes longer. Stir every 30 seconds.

MICRO-TIP: Exact cooking time will depend upon temperature of liquid. If one cup of drippings isn't available, milk and/or water may be used.

Rich Mushroom Sauce

1-1/2 cups of marvelous mushrooms.

3 tablespoons butter

1. Melt butter in 1-quart measure in Radarange Oven 1 minute.

3/4 cup half and half
1 (4 oz.) can drained mushrooms
1-1/2 tablespoons all-purpose flour
1 teaspoon soy sauce
1/4 teaspoon salt

2. Blend in half and half, mushrooms, flour, soy sauce and salt. Cook in Radarange Oven 2-1/2 minutes. Stir every 30 seconds.

Butter Sauce

1 cup of golden butter sauce.

1/2 cup butter

1. Melt butter in 1-quart measure for 1 minute in Radarange Oven.

1 cup sugar
3/4 cup light cream

2. Mix in sugar and cream. Cook in Radarange Oven 3-1/2 to 4 minutes. Stir every minute.

Barbecue Bean Sauce For Franks

6 to 8 servings of beans to bellow about.

1 (16 oz.) can barbecue beans
1/4 cup bottled chili sauce
1/4 cup shredded Cheddar cheese
3 tablespoons drained pickle relish
1/4 teaspoon chili powder

1. Mix all ingredients in 1-1/2-quart casserole. Heat in Radarange Oven 3 minutes. Stir every minute.

Apple-Raisin Sauce

1-1/2 cups sauce extraordinaire!

2 tablespoons brown sugar
1 tablespoon cornstarch
1/8 teaspoon salt
1/8 teaspoon allspice

1. Combine brown sugar, cornstarch, salt and allspice in 1-quart casserole.

1 cup apple juice

2. Blend in apple juice. Cook in Radarange Oven 2-1/2 minutes. Stir halfway through cooking time.

1/4 cup seedless raisins

3. Mix in raisins. Cook 2 minutes. Stir every 30 seconds.

1/2 cup diced apples

4. Add apples just before serving.

Cranberry Sauce

2 cups of festive cranberry sauce.

1 lb. cranberries

1. Wash and place cranberries in 3-quart casserole.

2 cups sugar
1 cup water

2. Stir in sugar and water. Cook in Radarange Oven 8 minutes or until mixture has boiled. Allow berries to cool.

MICRO-TIP: To make a tasty variation, substitute orange juice for water.

Quick Lasagna Sauce

Little bit of Italy in 3 cups.

1/2 lb. hot Italian sausage
2 tablespoons olive oil

1. Remove meat from sausage casing. Cut into small slices. Preheat 9-1/2-inch Amana Browning Skillet 4-1/2 minutes. Brown sausage in oil 2 minutes in Radarange Oven.

1/2 cup chopped onion
1/4 cup chopped green pepper
1 minced clove garlic

2. Thoroughly mix onion, green pepper and garlic with sausage. Cook in Radarange Oven 2 minutes. Drain all but 1 tablespoon of fat from skillet.

2 (10-1/2 oz. each) cans tomato sauce
1/2 teaspoon oregano
1/4 teaspoon basil
Salt
Pepper

3. Stir in tomato sauce and spices. Cook in Radarange Oven 3 to 4 minutes. Stir halfway through cooking time.

Strawberry Preserves

3 cups of sweet preserves.

1 (16 oz.) pkg. frozen strawberries

1. Place strawberries in 2-quart casserole. Cook in Radarange Oven 2 minutes.

3 tablespoons powdered fruit pectin

2. Stir in powdered fruit pectin. Cook 2 minutes, or until a few bubbles surface.

2 cups granulated sugar
1 tablespoon lemon juice

3. Stir in sugar and lemon juice. Cook in Radarange Oven 6 minutes, stirring every 2 minutes. Pour into glass jars. Cover, and refrigerate.

MICRO-TIP: If sealed with paraffin, jars may be stored in the freezer for use when needed.

Peach Conserve

10 (6 oz.) glasses or jars of peach happiness.

2 cups chopped peaches
1 (1 lb. 4 oz.) can crushed, drained pineapple
1 (8 oz.) bottle chopped maraschino cherries
Maraschino cherry juice

1. Combine peaches, pineapple, cherries and juice in 4-quart casserole. Cook, covered, in Radarange Oven 10 minutes.

7-1/2 cups sugar

2. Stir in sugar. Return to Radarange Oven, uncovered, 15 minutes. Stir every 5 minutes. Mixture should be boiling.

1 (6 oz.) bottle liquid fruit pectin

3. Stir in liquid fruit pectin. Cook 10 to 15 minutes longer. Temperature should read 228° F. Stir 5 minutes to cool. Pour into jars or glasses. Seal with paraffin.

Grape Jelly

2-4 (8 oz.) glasses of "jiggly" jelly.

6 cups Concord grapes
2 cups diced apples
1 cup water

1. Combine grapes, apples and water in 4-quart casserole. Cook, covered, in Radarange Oven 15 to 20 minutes or until fruit is tender. Stir every 5 minutes.

3/4 cup sugar for each cup juice

2. Strain juice, using several thicknesses of cheese cloth or jelly bag. Measure juice. Return to Radarange Oven. Bring to boil, about 10 minutes. Add sugar. Stir well. Cook to 220° F., about 30 minutes. Stir at 5-minute intervals. Pour into glasses to set. Seal with paraffin.

163
Desserts

The Radarange Oven is a "natural" for preparation of many desserts, ranging from simple applesauces to fruit crisps, puddings and custards.

Puddings will become popular desserts after you try them in the Radarange Oven. You'll find they are fun to prepare in your glass measuring cups.

Custards can also be made with ease. Just remember not to overcook them. If overcooked, they will curdle and separate. Individual custards can be set in a glass baking dish containing about an inch of water to help equalize the cooking.

In this chapter, you'll find recipes for the above mentioned desserts, as well as recipes for impressive cheesecakes and fruit toppings. All desserts can be prepared and cooked in minutes in the Radarange Oven!

Fruit Surprise Pudding

A delightful fancy fruit pudding dessert for 8.

1/3 cup butter or margarine
1 cup sugar

1. Cream butter. Beat in sugar gradually until light and fluffy.

2-1/2 cups all-purpose flour
2 teaspoons baking powder
1/2 teaspoon salt

2. Mix and sift dry ingredients.

1 cup milk
1 teaspoon vanilla

3. Combine milk and vanilla. Stir in dry ingredients alternately with milk to sugar mixture.

1/2 cup chopped, drained maraschino cherries
1 (8 oz.) can drained, crushed pineapple
1 tablespoon butter or margarine

4. Spread batter in sprayed 2-quart baking dish. Sprinkle with cherries and pineapple. Dot with butter.

2 cups firmly packed brown sugar
1-1/2 cups boiling water

5. Spread brown sugar in even layer over fruit. Pour boiling water over all.
6. Cover with waxed paper. Cook in Radarange Oven 10 to 12 minutes. Turn dish every 2-1/2 minutes. Let stand, covered, 10 minutes before serving.

MICRO-TIP: May be served warm with whipped cream.

Norwegian Prune Dessert

6 servings of a Norwegian favorite.

1-1/2 cups water
1/2 lb. pitted prunes

1. Boil water in 1-1/2-quart covered casserole 4 minutes in Radarange Oven. Stir in prunes. Let stand, covered, 2 hours.

1/2 cup sugar
1/2 cup chopped walnuts
1 tablespoon lemon juice
1/2 teaspoon cinnamon
2 tablespoons cornstarch

2. Drain prunes from liquid. Stir sugar, walnuts, lemon juice, cinnamon and cornstarch into liquid.
3. Cook above mixture in Radarange Oven 2-1/2 minutes. Stir well halfway through cooking time.

1/4 cup sherry

4. Blend in wine and prunes. Chill.

MICRO-TIP: May be served in sherbet glasses and topped with almond-flavor whipped cream.

ruit Surprise Pudding

Apple Crisp

A wisp of apple crisp — 6 servings.

5 cups sliced apples

3/4 cup quick-cooking rolled oats
1 cup all-purpose flour
1 cup firmly packed brown sugar
1/2 teaspoon salt
1 teaspoon cinnamon

1/2 cup butter or margarine

1. Place apple slices in 2-quart glass utility dish.
2. Combine oats, flour, brown sugar, salt and cinnamon.
3. Cut in butter until mixture is crumbly. Sprinkle evenly over sliced apples.
4. Bake in Radarange Oven 15 minutes, turning dish quarter-turn every 2 minutes.

Rhubarb Crisp

Reach for Rhubarb Crisp to serve 8.

4 cups, 1/2-inch pieces, rhubarb
1/2 cup sugar
2 tablespoons lemon juice
1/2 teaspoon grated lemon peel

1 cup firmly packed brown sugar
1 cup all-purpose flour
1/4 lb. soft butter or margarine

1. Combine rhubarb, sugar, lemon juice and lemon peel. Spread in bottom of 8 x 8 x 2-inch glass dish.
2. Mix together brown sugar, flour, and butter until crumbly. Sprinkle over rhubarb.
3. Bake in Radarange Oven about 15 minutes, or until rhubarb is tender. Turn dish quarter-turn every 2 minutes.

MICRO-TIP: May be served warm or cold with whipped cream or vanilla ice cream.

Cherry Crunch Dessert

Cherry Crunch for your bunch of 6.

1-1/2 cups sifted all-purpose flour
3/4 cup quick-cooking rolled oats
1 cup firmly packed brown sugar
1/2 teaspoon soda
1/2 teaspoon salt

1/2 cup softened butter

1 (1 lb. 5 oz.) can prepared cherry pie filling

1. Combine flour, oats, brown sugar, soda and salt in large bowl. Mix well.
2. Cut in butter and blend until particles are small and uniform. Pat half of crumb mixture into 9 x 9 x 2-inch glass baking dish.
3. Cover crumb layer evenly with pie filling. Sprinkle remaining crumb mixture over filling.
4. Bake in Radarange Oven about 15 minutes. Turn dish every 5 minutes. Serve warm or cold.

Custard Sauce

2-1/4 cups of creamy custard.

- 4 egg yolks
- 1/3 cup sugar
- Salt

- 1-1/2 cups milk
- 1/2 cup evaporated milk
- 1 teaspoon vanilla

1. Beat egg yolks. Mix in sugar and salt until well-blended.
2. Stir in milk, evaporated milk, and vanilla.
3. Cook in Radarange Oven 2-1/2 minutes, stirring at 30-second intervals. Chill before serving.

MICRO-TIP: Variations of this sauce are possible.
Rum Custard Sauce: Follow above recipe. Reduce vanilla to 1/2 teaspoon. Stir in one teaspoon rum flavoring.
Sherry Custard Sauce: Follow above recipe. Reduce vanilla to 1/2 teaspoon. Stir in one teaspoon sherry flavoring.
Lemon Custard Sauce: Follow above recipe. Stir in teaspoon lemon flavoring and 1/2 teaspoon grated lemon peel.

Mocha Bread Custard

Different dessert to serve 6 chocolate-lovers

- 1/2 cup semi-sweet chocolate morsels

- 1 cup strong coffee
- 1 cup milk
- 1 tablespoon melted butter or margarine

- 3 beaten eggs
- 1/4 cup sugar
- 1/4 teaspoon salt
- 1/2 teaspoon vanilla

- 1-1/2 cups white bread cubes
- 1/4 teaspoon cinnamon

1. Place chocolate morsels in bowl. Cover, and heat in Radarange Oven 2 minutes until chocolate softens.
2. Add coffee, milk, and butter. Cook in Radarange Oven 3 minutes.
3. Mix eggs, sugar, salt and vanilla. Stir hot chocolate mixture into egg mixture.
4. Place bread cubes in 1-1/2-quart casserole. Pour chocolate-egg mixture on cubes. Sprinkle with cinnamon.
5. Set casserole in large glass dish. Add 1 cup boiling water to large dish. Cook in Radarange Oven 9 to 10 minutes. Turn dish every 3 minutes. Let stand 10 minutes.

MICRO-TIP: May be served warm or cold with whipped cream.

Pudding from Prepared Mix

Convenient, speedy dessert to serve 4.

1 (3-1/2 oz.) pkg. prepared pudding mix
2 cups milk

1. Turn package of pudding into 1-quart glass casserole. Stir in enough milk to dissolve pudding. Stir in remaining milk.
2. Cook 2 minutes in Radarange Oven. Stir.
3. Cook 3 to 3-1/2 minutes longer, until pudding is thickened. Stir eve 30 seconds.

Minute Tapioca Pudding

Serve 6 a tempting tapioca treat.

3 tablespoons minute tapioca
5 tablespoons sugar
1/8 teaspoon salt
2 cups milk
1 egg yolk

1. Mix tapioca, 3 tablespoons sugar, salt, milk and egg yolk into 1-1/2-quart glass casserole. Let stand 5 minutes.

1/2 teaspoon vanilla

2. Cook in Radarange Oven about 3 minutes. Stir, then cook about 2-1/2 to 3 additional minutes, until mixture comes to full boil. Stir in vanilla.

1 egg white

3. Let mixture stand at room temperature while preparing egg white mixture. Beat egg white until foamy, gradually beating in remaining 2 tablespoons sugar. Whip to soft peaks.
4. Fold egg white into hot mixture. Serve warm or chilled.

Vanilla Cream Pudding

Yes, Virginia! There is vanilla cream pudding; 4 servings.

2 cups milk

1. Put milk in glass bowl and scald, about 2-1/2 minutes in Radarange Oven.

1/4 cup cornstarch
1/3 cup sugar
1/2 teaspoon salt

2. Blend together cornstarch, sugar and salt.
3. Remove milk and blend in cornstarch mixture gradually with wire whip. Return to Radarange Oven and cook about 2 to 3 minutes. Stir every minute.

3 egg yolks

4. Beat egg yolks and blend in. Return to Radarange Oven for about 30 seconds.

2 tablespoons butter
1/2 teaspoon vanilla

5. Beat well for smooth consistency. Mix in butter and vanilla. Blend well. Pour into 4 custard cups or sauce dishes.

Spiced Fruit Compote

Captivate your guests with this compote for 4 to 6.

2 medium-size apples
1 (1 lb.) can pear halves

1 tablespoon pear syrup
1/2 cup whole cranberry sauce
1/8 teaspoon cinnamon
Dash cloves
Dash allspice

1. Peel apples and cut into eighths. Cut pear halves in half, lengthwise. Layer apples and pears in 1-1/2-quart glass casserole.
2. Combine pear syrup with whole cranberry sauce and spices. Spoon over apples and pears.
3. Cook, uncovered, about 9 to 12 minutes in Radarange Oven, until apples are tender. Serve warm or chilled.

   MICRO-TIP: One (1 lb.) can of either pineapple chunks or apricot halves may be substituted for the pears.

Pear Streusel A La Mode

6-8 slices of streusel.

1 (1 lb.) can drained pear halves

12 finely crushed gingersnaps
4 tablespoons sugar
4 tablespoons soft butter or margarine

Vanilla ice cream

1. Place pears in 9-inch round glass dish.
2. Combine gingersnap crumbs and sugar in small bowl. Blend well. Mix in butter and work with fingers until mixture is crumbly. Sprinkle over pears.
3. Bake in Radarange Oven about 4 minutes. Turn dish halfway through cooking time. Serve warm with scoop of vanilla ice cream.

Fruit Soup

10, one-cup servings of flavorful fruit soup.

1/2 lemon
1 (12-oz.) pkg. dried apricots
1/2 cup raisins
2 quarts water

2 tablespoons quick-cooking tapioca
1/2 teaspoon salt
1 cup sugar
3 cinnamon sticks

1. Thinly slice lemon. Combine mixed fruits, apricots, raisins, lemon slices and water.
2. Stir in tapioca, salt, sugar and cinnamon sticks.
3. Cook in Radarange Oven about 15 minutes or until boiling. Simmer using Slo-Cook Cycle 25 minutes, or until fruits are tender. Serve hot or cold.

Marvelous Marble Cheesecake

Guests will marvel at this cheesecake.

1 cup crushed vanilla wafers
1/4 cup sugar

1. Combine wafer crumbs and 1/4 cup sugar in 9-inch glass pie plate.

1/4 cup butter

2. Melt butter in Radarange Oven about 2 minutes. Blend butter with crumbs and sugar. Press evenly over sides and bottom of pie plate. Bake in Radarange Oven about 1 minute, 45 seconds. Turn every 20 seconds. Cool.

4 (3 oz. each) pkgs. cream cheese
2 eggs
1/2 cup sugar
1 teaspoon vanilla

3. Put cream cheese, eggs, 1/2 cup sugar and vanilla in mixing bowl. Beat with electric mixer until smooth and creamy. Turn into crumb crust.

1/3 cup semi-sweet chocolate or butterscotch morsels

4. Place chocolate in small glass dish. Melt in Radarange Oven about 2 minutes. Spoon melted candy over cheese filling. Using fork, ligh swirl candy into filling mixture, using minimum number of strokes.
5. Bake in Radarange Oven about 4 minutes, until outer edge of filling is set. Turn every minute. Cool to room temperature. Chill in refrigerator before serving.

Cheer-leader's Cheesecake

Cheer for cheesecake!

2 cups graham cracker crumbs
1/2 cup sugar
1/2 cup butter
1/2 teaspoon cinnamon

1. Combine crumbs, sugar, butter and cinnamon in 13 x 9 x 2-inch dish. Pat firmly onto bottom of dish. Cook in Radarange Oven 2 minutes.

3 (8 oz. each) pkgs. cream cheese
5 eggs
1 cup sugar
1/2 teaspoon vanilla

2. Beat cream cheese until smooth. Stir in eggs, one at a time. Mix in sugar and vanilla. Pour into crust. Bake in Radarange Oven 15 minutes. Turn every 4 minutes.

2-1/2 cups sour cream
1-1/2 teaspoons vanilla
1/3 cup sugar

3. Combine sour cream, sugar and vanilla. Blend well. Pour mixture over cheesecake. Return to Radarange Oven 1 minute, 15 seconds. Chill.

MICRO-TIP: Fruit pie mixes may be spread over top.

Cranberry Whip Dessert

Crimson cranberry dessert to add color to a meal for 8.

1 cup sugar
1 cup water

2 cups fresh cranberries

1 (3 oz.) pkg. orange flavor gelatin

1/2 cup evaporated milk
1 teaspoon lemon juice

1. Combine sugar and water in 3-quart glass casserole. Bring to boil in Radarange Oven.
2. Stir in cranberries. Return to boiling point in Radarange Oven, about 6 minutes. Heat until berries pop open, about 6 more minutes.
3. Drain cranberries and reserve liquid. Add enough hot water to reserved liquid to make 2 cups.
4. Dissolve gelatin in hot liquid. Stir in cranberries and chill until thickened.
5. Chill evaporated milk in freezer until ice crystals form. Whip milk until frothy. Stir in lemon juice. Beat until stiff.
6. Fold whipped milk into thickened gelatin mixture. Pour into 10 x 10 x 2-inch glass dish. Chill.

Pineapple Bridge Dessert

Novel dessert for 9.

2/3 cup butter or margarine

1-1/2 cups vanilla wafer crumbs

2-1/2 cups crushed pineapple
1 (3 oz.) pkg. lemon flavor gelatin

1/2 cup sugar
3 eggs
1/2 cup chopped nuts

1/4 cup sugar

1. Melt half of butter in 8 x 8 x 2-inch glass baking dish for 30 seconds in Radarange Oven.
2. Combine wafer crumbs with butter. Line bottom of dish with 1 cup crumb mixture. Reserve remaining crumbs for topping.
3. Drain pineapple. Heat pineapple syrup in 1-1/2-quart glass casserole for about 2 minutes. Stir in gelatin and blend until dissolved. Cool to room temperature.
4. Cream remaining butter with 1/2 cup sugar. Separate eggs. Mix in egg yolks, one at a time, and beat well. Stir in cooled gelatin mixture, pineapple and nuts.
5. Beat egg whites until they form soft peaks. Gradually mix in 1/4 cup sugar and continue to beat until stiff.
6. Fold egg whites into pineapple mixture. Pour into crumb-lined dish and top with remaining crumbs. Chill until firm.

New England Carrot Pudding

Quaint carrot pudding for 8.

1/2 cup margarine
1/2 cup firmly packed brown sugar
1 egg
1 cup firmly packed grated raw carrots
2 teaspoons candied ginger
1/2 cup seedless raisins
1 cup currants

1. Cream margarine and brown sugar. Beat in egg until well-blended. Stir in carrots, ginger, raisins and currants.

1-1/4 cups cake flour
1 teaspoon baking powder
1/2 teaspoon soda
1/2 teaspoon cinnamon
1/2 teaspoon nutmeg

2. Sift together flour, baking powder, soda, cinnamon and nutmeg. Stir into above mixture until well-blended.
3. Turn into well-greased 9 x 9 x 2-inch glass dish. Bake in Radarange Oven about 11 to 11-1/2 minutes, or until slightly firm throughout. Turn dish every minute. Cut pudding into squares while still warm.

MICRO-TIP: May be served with hard sauce or thin lemon sauce.

Date Nut Torte

Dynamite date-nut bars for 9.

1/4 cup all-purpose flour
1/2 teaspoon baking powder
1/4 teaspoon salt
1-1/2 cups chopped, pitted dates
1-1/2 cups chopped nuts

1. Sift together flour, baking powder and salt. Mix well with dates and nuts.

2 eggs
3/4 cup sugar
1/2 teaspoon vanilla

2. Separate eggs. Beat egg yolks until very thick. Mix in sugar gradually while continuing to beat until very thick and fluffy. Fold in date-nut mixture and vanilla.
3. Beat egg whites until stiff but not dry. Fold into above mixture.
4. Turn into well-greased 9 x 9 x 2-inch glass dish. Bake in Radarange Oven about 3 minutes. Turn dish quarter-turn. Continue baking about 8 minutes, turning dish every minute. Cool. Cut into squares.

MICRO-TIP: Torte may be served with whipped cream.

Fabulous Fruit Cake

2 loaves of fancy fruit cake.

- 1 lb. chopped pitted dates
- 1 lb. chopped mixed candied fruits
- 3 cups pecan halves

1. Combine dates, candied fruits, and pecans.

- 1 cup all-purpose flour
- 1/2 cup sugar
- 1 teaspoon baking powder
- 1/4 teaspoon salt
- 4 eggs
- 1 teaspoon vanilla

2. Sift dry ingredients together. Combine with fruit and nuts. Beat eggs well. Stir in eggs and vanilla. Mix thoroughly.
3. Turn batter into two 8-1/2 x 4-1/2 x 2-1/2-inch glass loaf dishes. Bake each loaf in Radarange Oven about 7 to 8 minutes. Turn ever minute.
4. Completely cool cake before turning out of dish.

MICRO-TIPS: (1) An electric knife may be used to slice cake. This will assure clean-cut pieces of fruit.

(2) Fruit cake is best when it's allowed to stand several weeks before using.

Fudge Pudding Cake

8 servings of popular pudding cake.

- 2 tablespoons butter or margarine
- 1/2 cup sugar
- 1 teaspoon vanilla

1. Melt butter or margarine in medium-size mixing bowl in Radarange Oven. Combine with 1/2 cup sugar and vanilla.

- 1 cup all-purpose flour
- 3 tablespoons cocoa
- 1 teaspoon baking powder
- 1/2 teaspoon salt
- 1/2 cup milk
- 1/2 cup chopped nuts

2. Mix flour, 3 tablespoons cocoa, baking powder, and 1/2 teaspoon salt. Stir this mixture into first mixture, alternately with milk. Stir in nuts.

- 1-2/3 cups boiling water
- 1/2 cup sugar
- 5 tablespoons cocoa
- 1/4 teaspoon salt

3. Combine boiling water with remaining sugar, cocoa and salt in 8 x 8 2-inch glass dish. Drop batter from Step 2 by rounded tablespoons onto this boiling mixture.

- Confectioners' sugar

4. Place in Radarange Oven about 8 minutes. Turn after 5 minutes. Sprinkle with confectioners' sugar. While still warm, spoon out portions of cake.

MICRO-TIP: May be served with a sauce or whipped cream, if desired.

Fruit Sauce

2 cups of a colorful combination!

1 (10 oz.) pkg. frozen, sliced strawberries
1/4 cup sugar
1 tablespoon cornstarch

1. Defrost, blend berries, sugar and cornstarch in 1-quart glass measure. Cook in Radarange Oven 2 minutes. Stir mixture halfway through cooking time.

1/2 cup white rosé or fruit flavored wine

2. Stir in wine. Cook in Radarange Oven 1 minute. Stir mixture halfway through cooking time. Chill.

Melba Sauce

3 cups of infinite creativity.

1 (10 oz.) pkg. frozen raspberries

1. Place frozen raspberries in 1-quart casserole. Defrost in Radarange Oven 2 minutes.

1/2 cup sugar
2 tablespoons cornstarch

2. Combine sugar and cornstarch, and then stir into raspberries.

1/2 cup currant or apple jelly

3. Fold in jelly. Cook in Radarange Oven 3 minutes. Stir halfway through cooking time. Strain, if desired, and then cool.

MICRO-TIP: Especially good over ice cream.

Syrup for Crushed Ice Cone

8 crushed-ice cones to cool your day.

2 cups sugar
3/4 cup water

1. Combine sugar and water in medium-size glass bowl. Bring to rolling boil in Radarange Oven. Stir halfway through cooking time.

1 (0.15 oz.) envelope ade mix

2. Stir in ade mix until dissolved. Cool. Pour over finely chopped ice in paper cups. Use approximately 1/4 cup of syrup per cup of ice.

Fruit Flavored Topping

One generous cup of tangy topping.

1 cup water
1/2 cup sugar
1 (3 oz.) pkg. fruit flavor gelatin
2 tablespoons cornstarch

1. Combine ingredients in 2-cup glass measure. Stir. Cook in Radarange Oven 2 to 3 minutes. Let boil until thickened.

MICRO-TIPS: After removing metal lids, commercial ice cream toppings can be warmed in seconds in Radarange Oven.

Using paper plate, warm one cup of toasted almonds 2 minutes in Radarange Oven. Always a welcome treat on sundaes!

Bing Cherry Sauce

Brighten a dessert with 1-1/2 cups of bing cherry sauce.

1 (1 lb.) can bing cherries
3/4 cup cherry juice

1. Drain cherries and reserve juice. Measure juice and if necessary, add water to make 3/4 cup.

2 tablespoons sugar
1 tablespoon cornstarch
Dash salt

2. Combine sugar, cornstarch and salt in 1-quart glass sauce dish or measuring cup. Stir in reserved cherry juice.
3. Cook sauce in Radarange Oven until it is thick and clear, about 2 to 2-1/2 minutes. Stir after each minute.

1 teaspoon butter
1/2 teaspoon lemon juice

4. Stir in butter, lemon juice and cherries. Heat in Radarange Oven 45 seconds.

MICRO-TIP: May be served over cake, pudding or ice cream.

Hot Vanilla Sauce

1 cup vanilla sauce for variety.

3 tablespoons soft butter or margarine
1/2 cup sugar

1. Cream together butter and sugar in 1-quart glass casserole.

2 egg yolks
1/2 cup boiling water
Dash salt

2. Beat egg yolks slightly. Combine with butter and sugar mixture, then beat in water and salt.
3. Heat, uncovered, in Radarange Oven about 2 minutes. Stir well every 30 seconds. Be careful not to overcook.

1 teaspoon vanilla

4. Stir in vanilla. Serve sauce hot.

Dessert Lemon Sauce

1 cup of leprechaun's lemon sauce.

1/2 cup sugar
1 tablespoon cornstarch

1. Combine sugar and cornstarch in 1-quart casserole.

1 cup water

2. Stir in room-temperature water. Heat, uncovered, in Radarange Oven about 2 minutes. Stir every 30 seconds.

2 tablespoons butter or margarine
1/2 teaspoon grated lemon peel
1-1/2 tablespoons lemon juice
Dash salt

3. Blend in butter, lemon peel, lemon juice and salt.

MICRO-TIP: May be served warm or cold.

Cinnamon Apples

6 candied apples.

1 cup sugar
1 cup water
2 tablespoons red cinnamon drops

1. Combine sugar, water and red cinnamon drops in glass casserole. Cook about 3 minutes in Radarange Oven.

6 medium-size firm, tart apples
Whole cloves

2. Core and pare apples. Stick 3 or 4 whole cloves into each apple.
3. Place apples in the red cinnamon syrup. Cook about 7 to 8 minutes, or until apples are tender but firm. Turn apples every 2-1/2 minutes.
4. Cool apples in syrup before serving.

Baked Rhubarb Sauce

Refreshing rhubarb dish for 4.

2 cups chopped rhubarb
2 tablespoons water
Dash salt

1. Combine rhubarb and water in 2-quart glass casserole. Stir in salt.
2. Cover and bake in Radarange Oven about 4 minutes. Stir halfway through cooking time.

1/2 cup sugar

3. Mix in sugar and cook 1 additional minute. Keep rhubarb covered while cooling.

Baked Grapefruit

Once "hard to do it", now "nothing to it".

1 grapefruit

1. Cut grapefruit in half and cut around each section with sharp knife. Remove seeds.

2 teaspoons brown sugar
1 teaspoon butter

2. Sprinkle brown sugar over top. Dot with butter.
3. Place grapefruit on paper plate or in glass casserole. Bake in Radarange Oven, uncovered, for about 2-1/2 minutes. Serve hot.

Baked Apples

4 apples to applaud.

4 medium-size apples

1. Core apples and slice thin circle of peel from top of each apple. Arrange apples in 9-inch round glass dish.

1/4 cup sugar
Butter or margarine

2. Spoon 1 tablespoon sugar into each apple cavity. Place small piece of butter on each apple.
3. Cook about 4 to 5 minutes in Radarange Oven or until apples are tender. Let apples stand a few minutes before serving.

MICRO-TIP: Apples may be filled with mincemeat, whole cranberry sauce, or raisins and nuts. If this is done, increase cooking time by 1 or 2 minutes.

Cakes

You will be most pleased with cakes baked in the Radarange Oven because they are so moist and light. It is important to undercook them, rather than overcook, in order to prevent dehydration and toughness. Also, there is a greater amount of carry-over cooking time in microwave cooking, and you will want to take foods out at a more "underdone" stage than when conventionally preparing them. A slightly moist spot on the top of a cake does not necessarily mean the batter isn't fully cooked. After standing, the cooking process will be completed, and the moist spot will disappear.

Hints for Successful Cakes

Cakes may be baked in differently shaped utensils, and so cooking times will vary according to the dish shape selected.

Utensil	Radarange Time
8 or 9-inch round	5 minutes
8 x 8 x 2-inch square	7 minutes
9 x 13-inch oblong	10 minutes
Ceramic bundt dish*	12-14 minutes

*Note: if you do not have a ceramic bundt dish, a 3-quart mixing bowl may be substituted. Invert a 2-inch diameter glass in the center of the bowl.

•Cake mixes will rise higher when cooked in the Radarange Oven. Remember to fill the baking utensil only half full. Extra batter may be used for making cupcakes.

•Turning a cake several times during baking will assure a more evenly cooked cake.

•Test cakes with a toothpick or a touch of the finger.

•Refrigerate cakes a few minutes before frosting so frosting won't melt.

Cupcakes

Baked cupcakes should be taken out of the glass custard cups upon being removed from the microwave oven. Otherwise, a small amount of water will collect in the bottoms of the cups. This is due to condensation which occurs from steam as the cupcakes cool. This accumulation of water could cause the bottoms of the cupcakes to become soggy if they are left in the glasses to cool.

Fill the paper baking cups one-third full of batter.

1 cupcake	30 seconds
2 cupcakes	60 seconds
6 cupcakes	2-1/2-3 minutes

When 3 or more cupcakes are baked at one time, they should be placed in a circle in the oven to insure even cooking.

•Remember: timing is important. Overcooking results in a dehydrated, tough product.

ncy Pistachio Nut Cake

Orange Crumb Cake

"Orange" you going to try this? 12 to 16 slices.

2 cups all-purpose flour
1 cup sugar
1/2 cup butter

2 teaspoons cinnamon

2 teaspoons baking powder

2 eggs
1 cup diluted frozen orange juice

1. Mix flour, sugar and butter together until mixture is crumbly like cornmeal.
2. Combine 1 cup of crumbled mixture with cinnamon and reserve for topping.
3. Stir in baking powder with remainder of crumbled mixture and blend
4. Beat eggs with orange juice and mix lightly into crumb mixture.
5. Lightly grease 2-quart glass utility dish. Fill with cake batter. Spread evenly into corners. Sprinkle batter with cinnamon topping. Bake in Radarange Oven for about 8 minutes. Turn dish halfway through cooking time.

   MICRO-TIP: May be cut into squares and frozen to be later served for a quick brunch treat.

Butter Brickle Peach Cake

Crackling good cake: 12 to 16 slices.

1 (1 lb. 13 oz.) can sliced peaches

1 pkg. butter brickle cake mix
1/2 cup softened butter

1. Empty peaches and syrup into 2-quart glass utility dish.
2. Blend dry cake mix with butter. Sprinkle over peaches.
3. Bake in Radarange Oven 20 minutes. Turn dish every 5 minutes.

Tomato Spice Cake

Serve 12 a tantalizing tomato spice cake dessert.

1-3/4 cups all-purpose flour
3 teaspoons baking powder
1 cup sugar
1/2 teaspoon cinnamon
1/2 teaspoon cloves
1/2 teaspoon nutmeg

1. Sift dry ingredients together in mixer bowl.

1/2 cup soft shortening
1/4 cup water
1 (10-oz.) can condensed tomato soup
2 eggs
3/4 cup chopped raisins

2. Beat in shortening, water, and half of soup until smooth. Blend in remaining soup and eggs. Stir in raisins.
3. Pour batter into 9 x 9 x 2-inch glass dish sprayed with vegetable coating. Cover with plastic film.
4. Bake in Radarange Oven 12 minutes. Turn every 2-1/2 minutes. Let stand 10 minutes before serving.

MICRO-TIP: May be served with an orange glaze.

Rhubarb Cake

Enough for second servings. 12 to 16 slices.

1-1/2 cups firmly packed brown sugar
1/2 cup shortening
1 egg

1. Cream brown sugar and shortening. Stir in egg and beat well.

2 cups unsifted all-purpose flour
1 teaspoon soda
1 cup sour milk or buttermilk

2. Combine flour and soda. Mix into egg and shortening mixture alternately with buttermilk.

1-1/2 cups finely cut rhubarb
1 teaspoon vanilla

3. Fold in rhubarb. Stir in vanilla. Spread in 2-quart glass utility dish.

1/2 cup sugar
1 teaspoon cinnamon

4. Mix sugar and cinnamon. Sprinkle top with sugar and cinnamon mixture. Bake in Radarange Oven for about 20 minutes. Turn dish quarter-turn each minute for first 10 minutes. Turn dish half-turn halfway through last 10 minutes of cooking time.

Surprise Cupcakes

3 dozen surprises.

1 (17-1/2 oz.) pkg. chocolate cake mix
2 eggs
1-1/2 cups water

1. Mix cake mix with eggs and water according to package directions.

1 (8 oz.) pkg. cream cheese
1/3 cup sugar
1 egg
Pinch of salt
1 (6 oz.) pkg. semi-sweet chocolate morsels

2. Soften cream cheese. Cream together sugar and cheese. Mix in egg and salt. Beat until smooth. Fold in chocolate morsels.
3. Line 6 custard cups with paper bake cups. Drop batter by rounded teaspoon into each. Drop rounded teaspoon cream cheese mixture into center of cake batter. Cover by dropping another rounded tablespoon of chocolate mixture.
4. Bake 6 cupcakes in circle in Radarange Oven 3 minutes. Remove cupcakes from custard cups to cooling rack immediately. Repeat procedure for remaining cake batter.

MICRO-TIP: Cupcakes may be frosted with favorite icing.

Chocolate Icing Deluxe

Frosting to feature on 3 dozen cupcakes or a 2 layer cake.

1 large egg
2 cups confectioners' sugar

1. Beat egg with electric mixer until fluffy. Continue to beat while adding sugar gradually.

2 squares unsweetened chocolate

2. Melt chocolate in paper wrappers in Radarange Oven for about 1-1/2 to 2 minutes.

1/4 teaspoon salt
1/3 cup softened butter or margarine
1 teaspoon vanilla

3. Stir salt, shortening and chocolate into egg and sugar mixture. Beat until smooth and creamy. Stir in vanilla.

Cinnamon Coffee Cake

13 x 9 x 2-inch coffee or tea cake.

1-1/2 cups boiling water
1 cup quick-cooking rolled oats
1/2 cup butter

1. Pour water over oats. Break butter into chunks and drop onto oats. Stir mixture until butter melts.

1-1/2 cups all-purpose flour
1 teaspoon soda
1-1/2 teaspoons cinnamon
1 teaspoon salt

2. Sift flour, soda, cinnamon and salt. Stir into oat mixture. Mix well.

3/4 cup sugar
1 cup firmly packed brown sugar
2 eggs

3. Stir in sugar and brown sugar. Beat and stir in eggs. Mix thoroughly.

4. Bake in 2-quart utility dish about 12 minutes in Radarange Oven. Turn dish every minute.

3/4 cup firmly packed brown sugar
2 tablespoons milk
6 tablespoons butter

5. Combine sugar, milk and butter in 1-quart glass casserole. Bring to boil and continue cooking 1 minute in Radarange Oven.

1/2 cup chopped pecans
1 cup coconut

6. Blend in pecans and coconut. Spread on cake.

Banana Cake

2 layer banana cake.

1/2 cup butter or margarine
1-1/2 cups sugar

1. Cream shortening, stirring sugar in gradually.

3 eggs
1 cup mashed bananas

2. Separate eggs. Beat egg yolks and stir into sugar mixture. Mix bananas into sugar mixture.

1 teaspoon baking soda
1/2 cup commercial sour cream
2 cups all-purpose flour

3. Combine baking soda with sour cream. Stir sour cream mixture alternately with flour into banana mixture.

1 teaspoon vanilla

4. Beat egg whites until stiff. Fold in egg whites and vanilla.

5. Pour batter into two 9-inch round glass cake dishes. Bake each layer 6 minutes in Radarange Oven. Turn dish every minute. Cool.

Hot Frosted Gingerbread

A melt-in-your-mouth treat.

1/2 cup butter or margarine
1/2 cup hot strong coffee

1. Melt butter or margarine in hot coffee. If butter is not completely melted, place mixture in Radarange Oven for about 40 seconds, until melted.

2 eggs
1/2 cup sugar
1/2 cup molasses

2. Beat eggs and stir in sugar and molasses. Mix in coffee and butter.

1-1/2 cups all-purpose flour
2 teaspoons baking powder
1 teaspoon ginger

3. Sift flour, baking powder and ginger. Stir into first mixture. Combine ingredients well.

4. Turn into greased 8 x 8 x 2-inch glass dish. Bake in Radarange Oven about 8 minutes, turning dish quarter-turn every minute for first 5 minutes.

1 cup confectioners' sugar
2 tablespoons cream
1/2 teaspoon vanilla

5. Stir remaining ingredients together and spread on hot gingerbread.

Pineapple Upside-Down Cake

Hawaiian fruit cake fit for luau or lunch.

2 tablespoons butter

1. Place butter in 8 x 8 x 2-inch glass cake dish and melt in Radarange Oven 45 seconds.

1/2 cup firmly packed dark brown sugar

2. Blend brown sugar with butter and pack evenly into bottom of cake dish.

1 (15-1/4 oz.) can sliced pineapple

3. Drain pineapple juice into measuring cup. Arrange pineapple slices over butter-brown sugar mixture.

1 (9-1/2 oz.) pkg. one layer yellow cake mix

4. Prepare cake mix as directed on package, substituting pineapple juice for water. Turn cake batter into dish, distributing batter evenly over entire surface. Bake 7 minutes in Radarange Oven. Turn every minute.

Fancy Pistachio Nut Cake

You'll be "nuts" about this cake.

1 cup finely chopped pecans
3/4 cup sugar
2 tablespoons cinnamon

1. Generously grease ceramic bundt dish. Mix nuts, sugar and cinnamon in small bowl. Cover bottom and sides of dish with about 1/3 of nut mixture.

1 (1 lb. 2 oz.) box yellow cake mix
1 (3-1/2 oz.) pkg. instant pistachio pudding
4 eggs
1 cup sour cream
3/4 cup water or orange juice
1/4 cup oil
1 teaspoon vanilla

2. Blend cake mix, pudding, eggs, juice, sour cream, oil and vanilla. Alternate layers of batter with remaining nut mixture in greased dish. Swirl batter with fork. Bake in Radarange Oven 14 minutes, turning dish quarter-turn every 3 minutes. Cool 10 to 15 minutes before turning out on dish.

Radarange Coconut Cake

A charming white delight!

2-1/4 cups sugar
3/4 cup softened butter or margarine
3 beaten eggs
1-1/2 teaspoons vanilla

1. Gradually stir sugar into butter. Cream mixture. Mix in eggs and vanilla.

3-1/4 cups sifted cake flour
4 teaspoons baking powder
1-1/2 teaspoons salt

2. Sift flour, baking powder, and salt.

1-1/2 cups milk

3. Alternately mix dry ingredients and milk into creamed mixture. Begin and end with flour. Beat after each addition. Pour into three 9-inch round cake dishes. Bake in Radarange Oven 6 to 6-1/2 minutes per layer. Turn dish every minute. Cool.

1 (3-3/4 oz.) pkg. coconut or lemon cream pudding mix

4. Prepare pudding mix according to directions on package. Spread between layers of cake. Prepare frosting.

1-1/2 cups sugar
1/2 cup water
1/4 teaspoon cream of tartar

5. Combine sugar, water and cream of tartar in 1-1/2 -quart casserole. Cook, covered, in Radarange Oven bringing mixture to boil. Stir every minute. Uncover. Cook to hard ball stage, 266° F.

4 stiffly beaten egg whites
1 teaspoon vanilla

6. Gradually stir hot syrup, then vanilla into egg whites. Beat constantly.

1 cup shredded coconut

7. Frost cake. Sprinkle coconut on top and sides of cake.

Here are our favorite cookie and candy recipes. Bar-type cookies are quickly done in the Radarange Oven. The bars will not brown on the outside. However, granulated sugar mixed with cinnamon will enhance the color.

The Fruited Cookie Squares in this chapter make a nice treat with glazed fruit. The brown sugar in the bar gives a nice color.

Fruited Cookie Squares

16 fruited favorites.

1 cup diced fruit cake mix or chopped prunes, dates or other moist dried fruits
1 cup chopped walnuts or pecans
1 cup firmly packed brown sugar
3/4 cup all-purpose flour
1-1/2 teaspoons baking powder
1/4 teaspoon salt

1. Thoroughly mix first 6 ingredients.

3 eggs
2 tablespoons warm water
1 teaspoon vanilla

2. Beat eggs with water and vanilla. Stir into first mixture, blending thoroughly. Spread in 8 x 8 x 2-inch dish sprayed with vegetable coating. Cover loosely with plastic wrap.

Cinnamon
Sugar

3. Cook in Radarange Oven 8 to 8-1/2 minutes. Turn dish at 2 minute intervals. Let stand 10 minutes. Turn out on rack. Cut into squares. Sprinkle with cinnamon and sugar.

uited Cookie Squares

Chocolate Nut Balls

5-6 dozen chocolate chewies.

1 (5-1/3 oz.) can evaporated milk
1 (6 oz.) pkg. semi-sweet chocolate morsels

2-1/2 cups (12 oz.) crushed vanilla wafers
1/2 cup sifted confectioners' sugar
1-1/2 cups chopped walnuts
1/3 cup orange juice

1. In 2-quart glass dish, heat evaporated milk and chocolate morsels in Radarange Oven for about 2 minutes, until chocolate is melted. Stir halfway through cooking time. Mix until smooth.
2. Stir in crushed wafers, sugar, 1/2 cup of nutmeats and orange juice. Mix until well-blended. Let stand at room temperature for one hour.
3. Drop chocolate mixture by half-teaspoons into remaining nuts. Roll to coat with nuts and form balls. Refrigerate and store in closed container.

Unbaked Butterscotch Cookies

6 dozen butterscotch beauties.

2 cups sugar
3/4 cup butter or margarine
1 (6 oz.) can evaporated milk

1 (3-3/4 oz.) pkg. instant butterscotch pudding mix

3-1/2 cups quick-cooking rolled oats

1. Mix sugar, butter and milk in large bowl. Cook about 6 minutes in Radarange Oven, until boiling. Stir every 1-1/2 minutes.
2. Gradually stir in pudding mix until dissolved.
3. Stir in rolled oats. Mix thoroughly.
4. Let cool 20 to 30 minutes until cool enough to touch. Drop from teaspoon onto waxed paper.

Frosted Oatmeal Squares

4 dozen splendid squares.

4 cups quick-cooking rolled oats
1 cup firmly packed brown sugar
1 cup soft margarine
1/2 cup white corn syrup

1 (6 oz.) pkg. semi-sweet chocolate morsels
3/4 cup chunk peanut butter

1. Mix oats, brown sugar, margarine and syrup thoroughly. Divide mixture between two 2-quart glass utility dishes. Press evenly into dishes and bake each 3-1/2 to 4 minutes in Radarange Oven until bubbly over entire top. Cool.
2. Melt chocolate morsels and peanut butter in 1-quart bowl for 1-1/2 minutes in Radarange Oven. Stir halfway through cooking time. Mix until smooth and spread on top of oatmeal mixture.
3. Store in refrigerator. The squares will seem hard to cut, but will soften at room temperature after a few minutes. Cut each dish into 24 squares.

Chewy Peanut Butter Bars

36 peanut butter-pleasers.

- 1/3 cup shortening
- 1/2 cup peanut butter
- 1/4 cup firmly packed brown sugar
- 1 cup sugar

- 1 teaspoon vanilla
- 2 eggs

- 1 cup all-purpose flour
- 1 teaspoon baking powder
- 1/4 teaspoon salt
- 1 (3-1/2 oz.) can flaked coconut

1. Cream together shortening, peanut butter, and sugars until light and fluffy.
2. Add vanilla and eggs. Beat well.
3. Mix in flour, baking powder and salt, stirring until thoroughly blended. Stir in coconut.
4. Spread evenly in greased 2-quart utility dish. Bake 8 minutes in Radarange Oven, turning dish quarter-turn every 2 minutes. Cool. Cut into bars.

Spicy Pumpkin Squares

2 dozen perfect pumpkin pleasures.

- 1/2 cup soft butter or margarine
- 1 cup firmly packed brown sugar
- 1 egg
- 1/2 cup canned pumpkin

- 1-1/2 cups unsifted all-purpose flour
- 1 teaspoon cinnamon
- 1/2 teaspoon ginger
- 1/2 teaspoon allspice
- 1/2 teaspoon soda
- 1/2 cup raisins
- 1/2 cup chopped walnuts or pecans

1. In large electric mixer bowl, cream together butter and brown sugar until fluffy. Beat in egg and pumpkin until combined. Mixture will look separated.
2. Sift flour with spices and soda. Carefully blend flour into creamed mixture. Stir in raisins and nuts.
3. Spread batter evenly in ungreased 2-quart utility dish. Place in Radarange Oven for 8 minutes, turning dish every minute for first 4 minutes of baking. When done, mixture will pull away from sides of dish. Remove to rack until cool, then spread on icing. When icing is set, cut into squares.

   MICRO-TIP: For special icing, blend 1 cup unsifted confectioner's sugar with 4 teaspoons orange juice concentrate, and 2 teaspoons milk or light cream, until smooth.

Applesauce Squares

2 dozen alluring applesauce squares.

1/2 cup soft butter or margarine
1 cup firmly packed brown sugar
1 egg
1/2 cup applesauce
Grated peel of 1 lemon

1. In large bowl, cream butter and sugar. Beat in egg, applesauce and lemon peel.

1-1/2 cups all-purpose flour
1 teaspoon cinnamon
1/4 teaspoon ground cloves
1/2 teaspoon soda
3/4 cup raisins or walnuts

2. Blend flour, spices and soda into creamed mixture. Stir in raisins or walnuts.
3. Spread batter evenly in ungreased 1-1/2-quart glass utility dish. Place in Radarange Oven 8 minutes, turning dish every minute for first 4 minutes.

   MICRO-TIP: Confectioners' sugar may be sifted over squares if desired.

Dream Bars

36 graham-chocolate goodies.

22 crushed (1 cup crumbs) graham crackers
1 (6 oz.) pkg. semi-sweet chocolate morsels
1 (3-1/2 oz.) can coconut
1 (15 oz.) can sweetened condensed milk

1. Combine all ingredients. Place in 8 x 8 x 2-inch glass dish.
2. Bake 8 minutes in Radarange Oven, turning every minute. Cool and cut into bars.

Crunchy Fudge Sandwiches

25 crispy-crunchy sandwiches.

1 (6 oz.) pkg. butterscotch morsels
1/2 cup peanut butter
4 cups crisp rice cereal

1. Melt butterscotch morsels and peanut butter in Radarange Oven 2 minutes. Add rice cereal and stir until well-coated.
2. Press half of cereal mixture into buttered 8 x 8 x 2-inch glass dish. Chill while preparing fudge mixture. Set remaining cereal mixture aside.

1 (6 oz.) pkg. semi-sweet chocolate morsels
1/2 cup sifted confectioners' sugar
2 tablespoons soft butter or margarine
1 tablespoon water

3. Combine chocolate morsels, sugar, butter and water in glass dish. Place in Radarange Oven 2 minutes to melt chocolate.
4. Spread fudge mixture over chilled cereal. Spread remaining cereal over top. Chill. Remove from refrigerator about 10 minutes before cutting into squares.

No-Fail Divinity

7 dozen pieces of dreamy divinity.

4 cups sugar
1 cup light corn syrup
3/4 cup water
Salt

1. Mix sugar, corn syrup, water and salt in 2-quart casserole. Cook in Radarange Oven 19 minutes, stirring every 5 minutes. Candy thermometer should read 260° F. If not, cook 1 or 2 minutes longer.

3 egg whites

2. While syrup cooks, beat egg whites very stiff in a large bowl.
3. Gradually pour hot syrup over egg whites and continue beating at a high speed until thick and candy starts to lose its gloss. Beating may require about 12 minutes.

1 teaspoon vanilla
1/2 cup nuts

4. Add vanilla and nuts to beaten mixture. Drop by teaspoons onto waxed paper.

MICRO-TIP: Candy may be tinted with food coloring for special occasions.

Radarange Quickie Fudge

Amana's favorite fudge.

1 lb. confectioners' sugar
1/2 cup cocoa
1/4 cup milk
1/4 lb. butter or margarine

1. Blend confectioners' sugar and cocoa in 8 x 8 x 2-inch dish. Pour in milk and place butter on top. Cook in Radarange Oven 2 minutes. Remove from Radarange Oven and stir just to mix ingredients.

1 tablespoon vanilla
1/2 cup chopped nuts

2. Add vanilla and nuts. Stir until blended. Place in freezer for 20 minutes or refrigerator for 1 hour. Cut and serve.

Old-Fashioned Taffy

1 pound of an old-fashioned, but old favorite taffy.

2 cups sugar
1/4 cup vinegar
1/4 cup water

1. Combine sugar, vinegar, and water in 1-1/2-quart casserole. Cook in Radarange Oven for 6 minutes. Stir well and cook for about 3 minutes. Remove and check with candy thermometer for 280°, soft crack.
2. Pour an equal amount of the mixture onto three well-buttered dinner plates. Allow to cool until at a temperature that feels warm to the hand but not hot.

1 teaspoon vanilla

3. Grease hands well and work in 1/3 teaspoon of vanilla for each plate. Pull until the taffy turns a shiny white color and is quite stiff to handle. Twist as a rope and snip the strand into bite-size pieces.

MICRO-TIP: If batches become too hard to pull, return plate to Radarange Oven to soften. Allow to cool, and pull the same as above.

Cinnamon Sugared Nuts

2 cups of coated nuts.

1/4 cup sugar
1/2 teaspoon cinnamon
2 tablespoons butter or margarine
2 cups pecan halves or walnut halves

1. Mix together sugar and cinnamon. Set aside.
2. Place butter in 10-inch ceramic skillet. Melt in Radarange Oven for 1 minute.
3. Combine nuts with butter in Radarange Oven for 5 minutes, until toasted. Stir every 1-1/2 minutes. Remove nuts from Radarange Oven and quickly stir in cinnamon-sugar mixture, coating nuts evenly.
4. Spread nuts in baking dish. Cool. Store in tightly covered container.

Caramel Apples

7-8 colossal caramel apples!

Wooden sticks
7-8 medium apples
1 lb. (2-1/2 cups) caramels
1 tablespoon water

1. Insert sticks into apples. Melt caramels with water in glass bowl in Radarange Oven 2 minutes. Stir until smooth.
2. Dip apples into melted caramels, twirling until apples are coated. Set on buttered waxed paper until cool.

MICRO-TIP: For 1 apple, melt 8 or 9 caramels with 1 teaspoon water.

Banana Boats

2 "Amana Bananas".

2 medium bananas
24 miniature marshmallows
1 (3/4 oz.) milk chocolate bar

1. Peel back upper sections of banana peels on inside curves of fruit. Leave peels attached at stem ends.
2. Scoop out some of pulp and fill cavities with half of marshmallows.
3. Break chocolate bar into squares and put half on top of marshmallows. Replace strip of peel.
4. Wrap bananas with plastic film, leaving both ends open. Heat in Radarange Oven 1 minute or until chocolate melts.

Some-Mores

Tasty treat for 1. You'll want to make "some more".

1 whole graham cracker
1 marshmallow
1/2 milk chocolate candy bar

1. Break graham cracker in half and place on paper plate. Top with chocolate, then marshmallow. Place in Radarange Oven for 30 seconds.
2. Remove from Radarange Oven. Top with second half of graham cracker.

Chewy Chocolate Log

1-1/4 pound of chewy chocolate.

- 2 cups round oat cereal
- 1 cup crisp rice cereal
- 1 cup chopped English walnuts
- 3/4 cup light corn syrup
- 1/4 cup sugar
- 1 (6 oz.) pkg. semi-sweet chocolate morsels

1. Cut top from 1-quart milk carton. Butter inside of carton. Combine cereal and walnuts in bowl. Set aside.
2. Combine corn syrup and sugar in 1-quart glass casserole. Cook 3 minutes in Radarange Oven, stirring after each minute, until boiling.
3. Remove from Radarange Oven. Stir in chocolate morsels. Blend until mixture is smooth. Pour over cereal mixture and toss to coat evenly.
4. Pack firmly into milk carton. Chill 1 to 2 hours. Loosen with spatula to slip log from carton. Slice to serve.

MICRO-TIP: These cookies pack easily for mailing.

Caramel Peanut Puffs

30 puff pieces.

- 1 (14 oz.) pkg. caramels
- 3 tablespoons water
- 30 large marshmallows
- 1-1/2 cups chopped peanuts

1. Melt caramels with water in glass dish in Radarange Oven 2-1/2 to 3 minutes. Stir at end of each minute.
2. Dip marshmallows into caramel syrup using toothpicks and roll to coat completely.
3. Roll coated marshmallows in chopped nuts and place on waxed paper. Let dry at room temperature.

MICRO-TIP: If syrup gets thick, stir in 1/2 teaspoon water and reheat in Radarange Oven 30 seconds.

Popcorn Peanut Bars

16 popular popcorn peanut bars.

- 1/2 cup sugar
- 1/2 cup light corn syrup
- 1/2 cup peanut butter
- 1/2 teaspoon vanilla
- 3 cups popped popcorn
- 1 cup salted Spanish peanuts

1. Combine sugar and corn syrup in medium size bowl. Heat to full boil in Radarange Oven, about 2 to 2-1/2 minutes.
2. Stir in peanut butter and vanilla until smooth.
3. Mix popcorn with peanuts. Pour peanut butter mixture over popcorn and peanuts. Stir to coat all particles.
4. Pat firmly into 8 x 8 x 2-inch glass dish. Cool and cut into 2-inch squares.

195
Pies

Making pies can be a real art. With some coordination of your efforts you'll find that pies are a real joy to make in the Radarange Oven. Yes, you can also make crusts. You'll enjoy preparing our graham cracker crust and gingersnap crust recipes. If crusts are a threat to you, there are many ready-to-bake frozen crusts and ready-to-roll pie crust mixes. Use your imagination for the filling: custard, gelatin, cream or fruit.

Pie Crust
Pastry crusts cooked in the Radarange Oven are light, tender and flaky. However, they may brown less than when baked conventionally. You may wish to add color by brushing the pie crust with vanilla or a glaze of beaten egg yolk or egg white. Yellow food coloring might also be added to crust. If you desire a darker crust, place it under the conventional broiler for a few minutes.

When baking pies in the Radarange Oven, our preferred method is to bake the bottom crust before adding the filling and the top crust. The filled pie is then returned to Radarange Oven to complete the baking process. Starting with a baked pie shell will help to prevent a soggy bottom crust in the finished product.

range Mallow Pie

Gingersnap Crumb Crust

A "melt-in-the-mouth" crust.

1-1/3 cups gingersnap crumbs
6 tablespoons butter or margarine

1. Mix butter and crumbs until well-blended in 9-inch glass pie plate. Press crumb mixture firmly against bottom and sides of pie plate.
2. Bake in Radarange Oven 1-1/2 minutes turning dish every 30 seconds. Cool on wire rack. Fill with desired filling.

Graham Cracker Crust

The familiar favorite.

1-1/4 cups graham cracker crumbs
1/4 cup soft butter or margarine
1/4 cup sugar

1. Mix all ingredients until well-blended in 9-inch glass pie plate. Press crumb mixture firmly against bottom and sides of pie plate.
2. Bake in Radarange Oven 1-1/4 minutes, turning dish every 15 seconds. Cool on rack. Fill with desired filling.

MICRO-TIP: An easy way to press cracker crumbs into place in 9-inch glass pie plate is to use an 8-inch pie plate.

8 or 9 inch Baked Pastry Shell

One single pastry shell.

1 cup all-purpose flour
1/2 cup shortening
1/2 teaspoon salt
4-5 tablespoons cold milk

1. Sift flour and salt together. Cut in shortening with pastry blender. Sprinkle 1 tablespoon milk over flour mixture. Gently toss with fork. Repeat until all is moistened.
2. Flatten on lightly floured board. Roll from center to edge until dough is 1/8-inch thick. Allow to rest before final shaping. Place in 9-inch pie plate. Trim and flute edge. Cover shell with paper towel. Top with 8-inch glass pie plate to keep flat and prevent shrinkage.
3. Bake in Radarange Oven 3 minutes turning every minute. Remove paper towel and plate. Cook 1-1/2 minutes longer. Cool before filling.

MICRO-TIP: Shell may also be baked by simply piercing the sides and bottom generously with fork. Bake in Radarange Oven 2-1/2 to 3 minutes, turning plate every 30 seconds.

Peach Praline Pie

A "peach" of a pie.

2-1/2 cups drained sliced peaches
1/4 cup sugar
1 tablespoon quick-cooking tapioca
1 tablespoon lemon juice

1. Combine peaches, sugar, tapioca and lemon juice.

1/2 cup sifted all-purpose flour
1/4 teaspoon salt
1/4 cup firmly packed brown sugar
1/4 cup butter

2. Mix flour, salt, brown sugar and butter with a fork until mixture is crumbly.

1 baked 9-inch pie shell
1/4 cup chopped nuts

3. Sprinkle 1/3 of flour mixture in bottom of baked shell. Cover with peach mixture. Sprinkle remaining flour mixture over top of peaches. Sprinkle nuts on top.
4. Cover with square of waxed paper. Cook in Radarange Oven 3-1/2 minutes or until peaches are cooked.

1-2-3 Apple Pie

An apple pie in a "twinkle of the eye".

1 (1 lb. 5 oz.) can apple pie filling
1 baked 9-inch pastry shell

1. Turn pie filling into baked pastry shell.

3/4 cup milk
1 cup dairy sour cream
1 (3-3/4 or 3-5/8 oz.) pkg. instant vanilla pudding mix

2. Slowly combine milk with sour cream. Mix well. Stir in pudding mix and beat according to package directions.
3. Pour pudding mixture over pie filling. Chill.

2 tablespoons sliced toasted almonds

4. Spread almonds on paper plate. Toast in Radarange Oven for about two minutes. Stir halfway through cooking time. Sprinkle almonds over top of pie.

Angel Food Pie

6 servings of the "perfect" pie.

3/4 cup sugar
4-1/2 tablespoons cornstarch

1. Mix together sugar and cornstarch.

1-1/2 cups water

2. Place water in 1-quart measuring cup in Radarange Oven 2 minutes until boiling. Remove. Stir in sugar and cornstarch mixture. Return to Radarange Oven 1-1/2 to 2 minutes, stirring every 30 seconds.

1/2 teaspoon salt
3 egg whites
3 tablespoons sugar
1 teaspoon vanilla

3. Add salt to egg whites and beat until stiff. Stir in sugar and vanilla. Beat until creamy.

Baked pie shell

4. Pour cornstarch mixture slowly over egg whites, beating while pouring. Cool. Fill pie shell. Chill 2 hours.

1/2 cup heavy cream
1/2 square grated, bitter chocolate
3 tablespoons chopped nuts

5. Whip cream and spread over top. Sprinkle chocolate and nuts over whipped cream.

MICRO-TIP: Whipped cream may be sweetened and flavored as desired.

Coconut Macaroon Pie

8 servings of marvelous macaroon pie!

Butter
All-purpose flour

1. Butter and flour a 9-inch glass pie plate. Set aside.

4 egg whites
1 cup sugar
1 teaspoon vanilla

2. In electric mixer, beat egg whites until stiff. Continue to beat egg whites while adding sugar, 1 tablespoon at a time. Stir in vanilla.

1 cup graham cracker crumbs
1 teaspoon baking powder
1/4 teaspoon salt
1/2 cup shredded coconut
1/2 cup chopped walnuts

3. Thoroughly mix graham cracker crumbs, baking powder, salt, coconut and walnuts. Fold into beaten egg white mixture.

4. Spread mixture in prepared glass pie plate, piling slightly around sides of plate. Bake in Radarange Oven 9-1/2 to 10 minutes, turning plate quarter-turn every 2 minutes. Cool.

MICRO-TIP: Just before serving, center of pie may be filled with whipped cream.

Lemon Crumb Pie

Mellow yellow pie for 6.

1-1/3 cups graham cracker crumbs
1/4 cup butter or margarine

3 eggs
Grated rind and juice of 1-1/2 lemons
1 (15 oz.) can sweetened condensed milk

1/8 teaspoon salt

1. Combine crumbs and butter until well-blended in 9-inch glass pie plate. Reserve 1/4 cup. Press remaining crumb mixture firmly against bottom and sides of pie plate.
2. Separate eggs and beat yolks until very thick. Stir in lemon juice, grated rind and condensed milk.
3. Beat egg whites until stiff. Add salt. Fold lemon mixture into egg whites.
4. Pour into pie shell and sprinkle reserved crumbs on top. Bake in Radarange Oven 5 minutes, turning dish at 1-1/2-minute intervals. Cool at room temperature. Chill before serving.

Lemon Meringue Pie

8 servings of meringue pie in minutes.

1 cup sugar
1-1/4 cups water
1 tablespoon butter

1/4 cup cornstarch
3 tablespoons cold water
6 tablespoons lemon juice
1 teaspoon grated lemon peel

3 egg yolks
2 tablespoons milk
Baked 8-inch pie shell

* * * *

3 egg whites
6 tablespoons sugar
1 teaspoon lemon juice

1. Combine sugar, water and butter. Heat about 3 minutes in Radarange Oven until sugar dissolves. Stir.
2. Mix cornstarch with cold water. Stir into butter and sugar mixture. Return to Radarange Oven for about 2 minutes, stirring every 30 seconds. Stir in lemon juice and lemon peel.
3. Beat egg yolks with milk. Slowly stir into cornstarch mixture. Return to Radarange Oven for about 2-1/2 minutes. Stir every 30 seconds. Cool. Pour into pie shell.
4. Beat egg whites stiff, but not dry. Mix in sugar gradually. Pour in lemon juice last. Spread over cooled filling, sealing edges of pastry.
5. Bake in Radarange Oven for about 3 minutes. Turn dish quarter-turn every 30 seconds.

Orange Mallow Pie

A 9-inch light luscious dessert.

32 large marshmallows or
3 cups miniature marshmallows
1 tablespoon grated orange peel
3/4 cup orange juice
2 tablespoons lemon juice

1. Combine marshmallows, orange peel and juices in 1-quart casserole. Place in Radarange Oven for about 2-1/2 minutes. Stir every 30 seconds until marshmallows are blended. Chill until partially set.

1-1/2 cups chilled whipping cream or 1 pkg. whipped topping

2. In chilled bowl, beat cream until stiff. Fold in marshmallow mixture.

Baked 9-inch pie shell

3. Pour into pie shell. Chill until set, about 4 hours.

MICRO-TIP: Chopped toasted almonds or shredded orange peel may be sprinkled on top if desired.

Chocolate Almond Pie

Fantastic 9-inch pie.

1 (3-3/4 oz.) milk chocolate bar with almonds or
6 (1/2 oz.) milk chocolate bars with almonds
15 large marshmallows
1/2 cup milk

1. Break chocolate bars in pieces. Combine in 1-quart casserole chocolate bars, marshmallows and milk. Cook in Radarange Oven 3 minutes, stirring every 30 seconds.
2. Remove from Radarange Oven and mix until smooth. Cool completely.

1 cup heavy cream
1 9-inch prepared graham cracker crust

3. Gently fold in whipped cream. Pour into prepared crust. Chill 3 hours.

Chocolate Pie Filling

Choose this chocolate pie.

1 baked 8-inch pastry shell or crumb crust

1. Prepare pastry shell.

1 (6 oz.) pkg. semi-sweet chocolate morsels
2 tablespoons sugar
3 tablespoons milk

2. Combine chocolate morsels, sugar and milk in 1-1/2-quart glass bowl. Heat in Radarange Oven 2 to 2-1/2 minutes. Stir every 45 seconds to melt chocolate and dissolve sugar. Cool to room temperature.

3 separated, room-temperature eggs

3. Beat egg yolks into chocolate mixture, one at a time. In another bowl, beat egg whites until moderately stiff peaks are formed.
4. Fold egg whites into chocolate. Pour filling into prepared crust. Chill until firm.

MICRO-TIP: May be served with whipped topping.

Chocolate Star Pie Shell

One 9-inch chocolate dream.

2/3 cup chocolate stars or
1/2 cup semi-sweet chocolate morsels
2 tablespoons butter or margarine

1. Melt chocolate with butter in Radarange Oven 2 minutes. Stir every 45 seconds.

2-2/3 cups flaked coconut
2/3 cup confectioners' sugar
2 tablespoons milk

2. Stir in coconut, sugar and milk. Mix well. Press onto bottom and sides of 9-inch greased pie plate. Chill until firm.

Banana Cream in Chocolate Shell

Filling for 9-inch chocolate dream.

Chocolate Star Pie Shell

1. Prepare Chocolate Star Pie Shell. Refrigerate.

1 (3-1/4 oz.) pkg. vanilla pudding and pie filling mix
1-3/4 cups milk

2. Pour pudding mix into 1-quart casserole. Gradually stir in milk. Cook in Radarange Oven 5-1/2 minutes, until pudding thickens. Stir every minute. Cover with waxed paper. Chill.

1-1/2 cups miniature marshmallows
1/2 cup heavy cream
2 sliced bananas

3. After cooling, mix in marshmallows. Whip cream. Blend into vanilla mixture. Slice bananas into chocolate shell. Pour filling over bananas. Chill several hours.

Pumpkin Pie

Traditional Pilgrim's Pumpkin Pie.

2 eggs
1 (1 lb.) can pumpkin
3/4 cup sugar
1/2 teaspoon salt
1 teaspoon cinnamon
1/2 teaspoon ginger
1/4 teaspoon cloves
1 cup evaporated milk

1. Mix all ingredients.

Unbaked pie shell

2. Pour into unbaked pie shell. Bake in Radarange Oven 20 minutes, turn every 2 minutes.

Frothy Pecan Pie

Frothy favorite for everyone.

1-1/2 cups pecan halves

1. Spread pecans on paper plate. Toast in Radarange Oven 2 minutes, stir halfway through cooking time.

1/4 cup soft butter or margarine
1/2 cup firmly packed brown sugar
1 cup light corn syrup
1/2 teaspoon vanilla

2. Cream butter and brown sugar until fluffy. Stir in corn syrup and vanilla. Beat well.

3 eggs

3. Stir in eggs. Beat until mixture is smooth and well-blended.

9-inch baked pastry shell

4. Spread pecans evenly in pastry shell. Pour on egg mixture. Bake in Radarange Oven 8 minutes until filling is set. Cool.

203
Freezer to Radarange Oven

Your Radarange oven and freezer can be indispensible companion appliances. This is true whether you buy most of your foods fully prepared from the freezer case, or whether you prepare them in your kitchen yourself. This chapter includes a selection of recipes, from appetizers to desserts, that illustrates techniques used when defrosting, heating or cooking foods from the freezer.

Microwave energy is attracted more to thawed moisture than frozen moisture. Therefore, care must be taken with some frozen foods so that the outside, which thaws first, does not become overcooked before the center can reach the desired serving temperature. A simple way to control and equalize the heating or cooking is to stir or rearrange the food. However, many frozen foods do not lend themselves to stirring or rearranging. With these foods, it is usually advisable to slow the cooking once the defrosting starts. This is accomplished automatically in the Radarange Oven when you use the Automatic Defrost Cycle. Without these automatic controls, it is necessary to occasionally turn the oven off and allow the heat on the edges to penetrate and defrost the center without continual cooking of the outside.

The amount of automatic cycled defrosting required can be further affected by the way the food is frozen. A thick food item takes more cycle time to defrost to the center than either a thin food item or a food that is separated into parts. Also, a mass of food with a center mound requires more time to heat than if the food has a depressed center.

Here are some suggestions to keep in mind when preparing foods for freezer storage.

PREPARATION: Slightly undercook the food since it will cook some more while being reheated. Package food according to its end use and container. If a food will be heated in a certain casserole or dish, be sure it is frozen in that size and shape. If you don't wish to use a dish for freezer storage, line the dish with foil. Arrange the food in the container and freeze. Once frozen, lift the food from the dish, using the foil extensions. Wrap the food tightly and return it to the freezer. When ready to use, just remove it from the foil and return it to the baking dish or casserole. When you want to store food in the serving dish, just wrap the total dish in foil or place it in a heavy plastic bag.

To keep pieces from freezing together, first arrange them separated on a tray and freeze. Then, store them together in a heavy plastic bag. Otherwise, place a double thickness of waxed paper between each portion you wish to keep separate while frozen.

WRAPPING: The key to maintaining quality during freezing is an airtight package that keeps air out while retaining the moisture and flavor of the food. Eliminate extra air in the package by molding the package close to the food. Materials that make airtight packages for freezing include: heavy duty aluminum foil, plastic freezer containers, plastic freezer bags and specially waxed paper packages.

STORAGE: Most items in this section will maintain their original quality for 2 or 3 months. After this time, there may be changes in taste and texture. Items stored at a temperature of even 0° F. maintain their quality better than items stored in a freezer with above 0° F. temperatures.

Keep the above thoughts in mind as you use the recipes on the following pages. You will soon discover how your Radarange Oven and your freezer can combine to make meal preparation easier and more enjoyable.

oman Noodles

Cheese Puffs

32 puffs to please a crowd.

1 cup shredded sharp Cheddar cheese
1 tablespoon all-purpose flour
1 teaspoon curry powder
1 teaspoon garlic salt
1 tablespoon dry sherry wine

1. Mix together cheese, flour and seasonings. Fold in wine.

2 egg whites

2. Beat egg whites until stiff. Fold into first mixture.
3. Drop by level teaspoon onto cookie sheet. Freeze. Transfer frozen cheese "drops" to plastic bag for longer freezer storage.
4. To serve, place 1 frozen cheese ball on each of 12 crisp crackers. Arrange crackers on 7-inch round paper plate. Heat in Radarange Oven 1-3/4 to 2 minutes before serving.

Walnut Bacon Crisps

12 crisps for a crunchy treat.

12 short bacon strips
12 walnut halves

1. Wrap bacon strips around walnut halves, fastening with toothpicks. Place in circle on double layer of paper towel. Cover with another paper towel.
2. Cook in Radarange Oven 3-1/2 to 4 minutes, or until bacon is browned and crisp. Freeze in airtight container.
3. Before serving, cook crisps 6-1/2 to 7-1/2 minutes in Radarange Oven.

Virginia Yam Casserole

8 servings of a great country casserole.

2 lbs. yams or sweet potatoes
1/4 teaspoon salt
1/4 cup water

1. Peel yams and slice 1/2-inch thick. Place salt and yams in 2-quart glass casserole. Add water. Cook, covered, in Radarange Oven 15 minutes. Stir halfway through cooking time. Drain.

2/3 cup firmly packed brown sugar
1 (1 lb.) can applesauce
2-1/2 tablespoons butter

2. Arrange half of yams in 1-1/2-quart glass casserole. Sprinkle with 4 tablespoons brown sugar. Spoon half of applesauce over yams. Dot with 1 tablespoon butter. Repeat layers. Use remaining brown sugar and butter to top casserole.
3. Cook in Radarange Oven 4 minutes. Cover casserole tightly and freeze
4. Before serving, heat casserole in Radarange Oven 12 to 15 minutes.

Cabbage Rolls

6 servings of two cabbage rolls each.

12 cabbage leaves
1 tablespoon water

1 lb. ground beef
6 to 8 ozs. pork sausage
1/3 cup chopped onion
1/2 cup instant rice
1 egg
1 (15 oz.) can tomato sauce

1 tablespoon brown sugar
1/2 teaspoon leaf basil

1. Place cabbage leaves and water in large casserole or bowl. Cook, covered, in Radarange Oven 8 minutes, or until softened. Set aside.
2. Combine beef, sausage, onion, rice and egg. Mix in 1/2 cup of tomato sauce. Divide mixture into 12 parts. Place 1 part at thick end of of each cabbage leaf. Fold in sides and end to enclose meat mixture. Fasten with toothpick. Arrange in 2-quart utility dish, forming two rows lengthwise in dish.
3. Combine remaining tomato sauce, brown sugar and basil. Spoon over rolls.
4. Cook, covered with waxed paper, in Radarange Oven 15 minutes, or until just about tender. Cool. Wrap tightly and freeze.
5. Cook frozen casserole, covered with waxed paper, in Radarange Oven 10 minutes. Then, cook in Radarange Oven using Slo Cook or Automatic Defrost Control 10 minutes. Rotate dish. Cook in Radarange Oven 12 minutes, using Slo Cook Cycle, or until heated through.

auerbraten Steak

This thick round steak thaws, marinates and cooks in one step and serves 6 to 8.

2-1/2 to 3-lb. frozen top round steak (about 1 inch thick)
1-1/2 cups water
1/2 cup red wine vinegar
1 sliced medium onion
1 teaspoon salt
1/2 teaspoon whole cloves
1/8 teaspoon pepper

6 to 8 gingersnaps
2 tablespoons brown sugar

1. Place frozen steak in 2-quart utility dish. Add remaining ingredients except gingersnaps and brown sugar.
2. Cook, covered with waxed paper, in Radarange Oven 15 minutes or until steaming hot. Then, cook in Radarange Oven using Slo Cook or Automatic Defrost Control 90 minutes, or until meat is fork tender. Turn steak over halfway through cooking time.
3. Strain cooking liquid into 2-cup measuring cup. Add gingersnaps and brown sugar, mixing with fork until smooth and thickened.
4. Cook in Radarange Oven 1-1/4 minutes or until boiling. Serve sauce over meat.

Glazed Chicken Wings

About 16 tasty morsels to have waiting in the freezer.

2 lbs. chicken wings

1. Separate wing sections at joint. Set aside tip section. (See MICRO-TIP.) Arrange wing pieces in 9-inch pie plate with small ends toward center.
2. Cook, covered with waxed paper, in Radarange Oven 7 minutes, or until no longer pink. Drain. Cool. Arrange in single layer in shallow dish. Freeze until firm. Store in heavy plastic bag.

1/4 cup currant jelly
1/4 cup catsup
1/2 teaspoon salt

3. Combine jelly, catsup and salt. Place frozen chicken wings in shallow glass dish. Pour glaze mixture evenly over wings. Cook, covered with waxed paper, in Radarange Oven 11 minutes, or until glazed and tender.

MICRO-TIP: The glaze mixture can be combined and frozen. Heat in Radarange Oven 1 minute, or until melted. The tip section of the wings may be boiled and the broth used as soup stock.

Crab Balls

18 savory balls to quickly reheat from the freezer.

1 (6-1/2 or 7 oz.) can crabmeat
1 egg
1/3 cup dry bread crumbs
3 tablespoons sour cream
2 tablespoons finely chopped onion
1 teaspoon prepared horseradish

1. Drain crabmeat. Combine with remaining ingredients except cornflake crumbs. Form mixture into about 18 balls, 1-inch in diameter.

3 tablespoons cornflake crumbs

2. Coat each with crumbs. Place in single layer in shallow dish. Freeze until firm. Store in heavy plastic bag.
3. Arrange frozen balls on glass plate. Cook in Radarange Oven 3-1/2 minutes, or until heated through.

Caraway Cheese Crackers

Cheese, ale and caraway combine for topping 36 crackers.

1-1/2 ozs. cream cheese
2 tablespoons butter
2 cups shredded Cheddar cheese
1/2 cup all-purpose flour
1/3 cup beer
1/2 teaspoon caraway seed
1/2 teaspoon dry mustard

1. Heat cream cheese and butter in bowl in Radarange Oven 30 seconds or until softened. Blend in remaining ingredients except crackers.
2. Cook in Radarange Oven 2-1/2 minutes, or until cheese is melted and mixture is smooth. Stir every minute. Cool. Form into 36 balls, about 1/2-inch in diameter. Freeze until firm. Store in heavy plastic bag.

36 crackers

3. Place each ball on cracker on plate. Heat 12 at a time in Radarange Oven 2-1/2 minutes, or until cheese is bubbly.

Stuffed Cabbage Rol

OUNCE SERVINGS
KEEP FROZEN
Rolls
in tomato sauce

Basic Tomato Sauce

Five pints of sauce make 5 different dishes.

1 cup chopped onion
2 minced cloves garlic
2 stalks chopped celery
2 tablespoons cooking oil

2 (28 oz. each) cans whole tomatoes
1 (12 oz.) can tomato paste
2 tablespoons dry parsley flakes
1 tablespoon sugar
1 tablespoon salt
1 teaspoon leaf basil
1/2 teaspoon leaf oregano
1/4 teaspoon pepper
1 bay leaf
1/2 cup red wine, if desired

1. Combine onions, garlic, celery and oil in 3-quart casserole. Cook, covered, in Radarange Oven 5 minutes. Stir halfway through cooking time.
2. Stir in remaining ingredients. Cook, covered, in Radarange Oven 30 minutes. Stir every 10 minutes. Cool. Remove bay leaf. Divide sauce among 5 one-pint freezer containers. Cover, and freeze.

Saucy Chili

Add beans and beef to the basic sauce for 4 hearty servings.

1 pint frozen basic tomato sauce (see above recipe)

1 lb. frozen ground beef

1 (15 oz.) can kidney beans
1-1/2 teaspoons chili powder
1/4 teaspoon salt

1. Heat frozen sauce in Radarange Oven 2 minutes, or until loosened from container. Set aside.
2. Heat frozen beef in 2-quart casserole in Radarange Oven 5 minutes. Break meat into small pieces. Cook in Radarange Oven 4 minutes, or until no longer pink. Drain fat.
3. Stir in sauce and remaining ingredients. Cook, covered, in Radarange Oven 10 minutes, or until bubbly. Stir every 3 minutes.

Saucy Chops And Rice

Rice and chops combine with the sauce for 5 simple, super servings.

1 pint frozen basic tomato sauce (see recipe top of page)

5 (1-3/4 lbs.) pork chops

1 cup long-grain rice
1-1/2 cups water
1/2 cup chopped green pepper
1 teaspoon salt

1. Heat frozen sauce in Radarange Oven 2 minutes, or until loosened from container. Set aside.
2. Preheat 9-1/2-inch Amana Browning Skillet in Radarange Oven 4-1/2 minutes. Add chops. Cook in Radarange Oven 5 minutes, turning chops over halfway through cooking time.
3. Add sauce and remaining ingredients to skillet. Cook, covered, in Radarange Oven 20 minutes, or until rice and chops are tender. Turn chops and stir halfway through cooking time.

Saucy Chicken

Turn the basic sauce into Chicken Cacciatore for 4 to 5.

1 pint frozen basic tomato sauce (p. 208)

1. Heat frozen sauce in Radarange Oven 2 minutes, or until loosened from container. Set aside.

2-1/2 to 3 lbs. cut-up frying chicken

2. Arrange chicken, skin-side-up, in 2-quart utility dish. Break partially thawed sauce into pieces and spoon over chicken.

Grated Parmesan cheese
3/4 teaspoon salt

3. Cook, covered with waxed paper, 25 minutes, or until chicken is tender. Spoon sauce over chicken about halfway through cooking time. Serve with cooked spaghetti. Sprinkle generously with Parmesan cheese. Salt to taste.

Saucy Meat Balls

Add meat balls and serve with spaghetti for 5 to 6 servings.

1 lb. ground beef
6 to 8 ozs. pork sausage
1/2 cup rolled oats
1/3 cup milk
1 egg
3/4 teaspoon salt
1/8 teaspoon pepper

1. Combine ground beef, sausage, oats, milk, egg, salt and pepper. Shape into about 20 meat balls, 1-1/2 inches in diameter. Arrange in 2-quart utility dish.
2. Cook, covered with waxed paper, in Radarange Oven 8 minutes, or until no longer pink. Rearrange meat balls about halfway through cooking time. Drain fat.

2 pints frozen basic tomato sauce (p. 208)

3. Heat frozen sauce in Radarange Oven 3 minutes, or until loosened from container. Break into chunks. Add to meat balls.
4. Cook, covered with waxed paper, in Radarange Oven 10 minutes or until hot and bubbly. Stir every 3 minutes. Serve over cooked spaghetti.

Saucy Beef 'N' Macaroni Hotdish

Combine the sauce with macaroni and beef for 4 to 5 hotdish servings.

1 pint frozen basic tomato sauce (p. 208)

1. Heat frozen sauce in Radarange Oven 2 minutes, or until loosened from container. Set aside.

1 lb. frozen ground beef

2. Heat frozen beef in 2-quart casserole in Radarange Oven 5 minutes. Break meat into small pieces. Cook in Radarange Oven 4 minutes, or until no longer pink. Drain fat.

1 cup uncooked macaroni
1 cup water
3/4 teaspoon salt

3. Stir in sauce and remaining ingredients. Cook, covered, in Radarange Oven 12 minutes, or until macaroni is just about tender. Stir every 3 minutes. Let stand few minutes to finish cooking.

MICRO-TIP: A can of drained corn or other favorite vegetable can be added with macaroni.

Lobster Tails

4 small servings that start with frozen lobster tails.

4 small frozen lobster tails (about 2 ozs. each)
1-1/2 cups water
1 slice lemon
1/2 bay leaf
1/2 teaspoon salt

1. Place frozen lobster tails in 1-quart casserole. Add remaining ingredients.

Melted butter
Lemon wedges

2. Cook, covered, in Radarange Oven 7 minutes, or until mixture is steaming hot. Let stand few minutes. Remove lobster. Cut through shells to remove meat. Serve with melted butter and lemon wedges.

Turkey Roll With Stuffing Balls

Turkey and stuffing for 6, plus leftover meat for the next day.

3-1/2-lbs. frozen turkey roast

1. Leave frozen roast in cooking bag package or place in plastic cooking bag. Fasten with rubber band. Place in 2-quart utility dish.
2. Cook in Radarange Oven 10 minutes. Turn roast over. Pierce small hole in top side of bag to allow steam to escape. Cook in Radarange Oven 10 minutes. Then, cook in Radarange Oven using Slo Cook or Automatic Defrost Cycle 25 minutes, or until meat thermometer registers 160° when inserted in center of roast. Set aside.

1-1/2 cups water
1/4 cup butter
1 (6 oz.) pkg. chicken-flavor, saucepan-type, stuffing mix
1 egg

3. Combine water and butter in bowl. Heat in Radarange Oven 3 to 4 minutes, or until boiling. Stir in stuffing mix, including seasonings. Mix in egg.
4. Remove turkey roast from cooking bag. Place skin-side-up in baking dish. Add cooking juices from bag. Spoon dressing in "balls" around roast.

1/4 cup apple jelly
1 teaspoon lemon juice

5. Heat jelly, lemon juice and paprika in Radarange Oven 1 minute, or until boiling. Brush on roast.
6. Cook, covered with waxed paper, in Radarange Oven 20 minutes, or until meat thermometer inserted in center registers 185° . Let stand 10 to 15 minutes before slicing.

Stuffed Chicken Breasts

Keep these 6 to 8 servings handy in the freezer.

4 whole halved chicken breasts

1. Skin and bone chicken breasts. Place each breast boned-side-up between sheets of plastic wrap. Flatten to 1/8 inch thickness using flat side of meat mallet or rolling pin.

8 slices thinly-sliced sandwich ham
2 ozs. bleu cheese
1/2 cup finely chopped fresh tomato
1/4 teaspoon poultry seasoning

2. Place equal portions of ham, cheese and tomato on flattened chicken breasts. Sprinkle each with poultry seasoning. Roll up each breast, tucking in ends and sealing ham and cheese mixture inside.

1/4 cup all-purpose flour
2 slightly-beaten eggs
1/2 teaspoon salt
2/3 cup corn flake crumbs

3. Coat rolls with flour. Combine eggs and salt. Dip rolls in egg mixture. Coat with crumbs. Wrap tightly and freeze.

4. Arrange frozen rolls on rack in 2-quart utility dish. Cook, covered with waxed paper, in Radarange Oven 18 minutes, or until done. Turn rolls about halfway through cooking time.

Chicken-Broccoli Bake

Turn leftover chicken into a company freezer casserole for 6.

2 (10 oz. each) pkgs. frozen broccoli spears
2 (10-3/4 oz. each) cans condensed cream of chicken soup
1/2 teaspoon curry powder
1 tablespoon lemon juice
12 slices cooked chicken or turkey

1. Cook broccoli in Radarange Oven 8 minutes, or until thawed. Combine soup, curry and lemon juice. Spread half in 2-quart utility dish. Arrange broccoli and chicken over soup. Top with remaining soup. Wrap tightly and freeze.

1 cup crushed salad croutons

2. Cook frozen casserole, covered with waxed paper, in Radarange Oven 10 minutes. Then, cook in Radarange Oven using Slo Cook or Automatic Defrost Cycle 20 minutes. Top with crushed croutons. Cook, uncovered, in Radarange Oven 15 minutes, or until heated through.

Twice Baked Potatoes

6 to 8 servings that taste as good as when freshly prepared.

6 medium potatoes

1. Pierce skin of potatoes with fork. Cook in Radarange Oven 17 minutes, or until just tender. Turn potatoes over halfway through cooking time. Let stand 10 minutes. Cut potatoes in half lengthwise. Scoop out pulp into bowl, being careful not to tear shells.

3 tablespoons butter
1 teaspoon salt
1 teaspoon chopped chives
1/8 teaspoon pepper
1/4 cup sour cream
2/3 cup milk
1/2 cup shredded cheese

2. Mash potatoes well. Beat in remaining ingredients except cheese. Spoon into potato shells. Top with cheese. Wrap tightly and freeze. Before serving, place frozen potatoes in 2-quart utility dish. Cook, covered with waxed paper, in Radarange Oven 10 minutes. Then, cook in Radarange Oven using Slo Cook or Automatic Defrost Cycle 7 minutes, or until heated through.

Veal Parmigiana

Another specialty to keep in the freezer for a quick meal for 6.

1-1/2 lbs. veal round steak
1 slightly-beaten egg
2 tablespoons milk
1 teaspoon salt
1/3 cup dry bread crumbs
1/4 cup grated Parmesan cheese

1. Cut steak into 6 serving pieces. Pound with meat mallet. Combine egg, milk and salt in shallow dish. Combine crumbs and cheese on waxed paper. Dip meat into egg mixture. Then, coat with crumbs.

2 tablespoons butter

2. Preheat 9-1/2-inch Amana Browning Skillet in Radarange Oven 4-1/2 minutes. Add 1 tablespoon butter and half of meat. Cook in Radarange Oven 5 minutes, turning meat halfway through cooking time. Remove meat and set aside. Reheat skillet in Radarange Oven 2 minutes. Add remaining tablespoon butter and meat. Cook in Radarange Oven 5 minutes, turning meat halfway through cooking time. Return all meat to skillet.

1 (8 oz.) can tomato sauce
1/4 teaspoon Italian seasoning
1/8 teaspoon instant minced garlic

3. Combine tomato sauce, seasoning and garlic. Spoon over meat. Cook, covered, in Radarange Oven 12 minutes, or until meat is tender. Cool. Wrap tightly and freeze.

16 slices Mozzarella cheese

4. Cook frozen meat mixture, covered, in Radarange Oven 10 minutes. Then cook in Radarange Oven using Slo Cook or Automatic Defrost Cycle 5 minutes. Top with cheese. Cook, covered, in Radarange Oven 3 to 4 minutes, or until cheese is melted.

Beef Bourguignon

A French classic for 6 to 8 that improves with freezing.

6 slices bacon
2 lbs. cut-up beef stew meat
1/4 cup all-purpose flour
1 teaspoon salt
1/4 teaspoon pepper
1 (10-1/2 oz.) can condensed beef consommé
1 cup Burgundy wine
1 bay leaf

1. Cook bacon in 3-quart ceramic casserole in Radarange Oven 6 minutes, or until crisp. Set aside bacon. Preheat 9-1/2-inch Amana Browning Skillet in Radarange Oven 4-1/2 minutes. Coat meat with mixture of flour, salt and pepper. Place meat and bacon drippings in skillet. Cook in Radarange Oven 4 minutes, until browned. Stir halfway through cooking time. Stir in any remaining flour, consommé wine and bay leaf.

1 pint fresh mushrooms
2 cups sliced carrots
1 (16 oz.) can drained, whole onions

2. Cook, covered, in Radarange Oven 7 minutes or until mixture boils. Then cook in Radarange Oven using Slo Cook or Automatic Defrost Cycle 60 minutes, or until tender. Remove bay leaf. Cool. Spoon into two 1-1/2 pint freezer containers. Freeze. Before serving, heat frozen meat mixture in Radarange Oven 4 minutes, or until loosened from container. Combine with mushrooms, carrots and onions in 3-quart casserole.

French bread

3. Cook, covered, in Radarange Oven 25 minutes, or until vegetables are tender. Stir every 5 minutes. Serve with French bread.

Roman Noodles

A peppy noodle side-dish for 4 to 5.

1/2 cup chopped onion
1 chopped green pepper
1/4 cup butter

1. Combine onion, green pepper and butter in 2-quart casserole. Cook in Radarange Oven 2 minutes. Stir every 30 seconds.

1 (4 oz.) can sliced mushrooms
2 (8 oz. each) cans tomato sauce
1/4 teaspoon powdered thyme
1/2 teaspoon salt
1/8 teaspoon pepper

2. Stir in mushrooms with liquid, tomato sauce and seasonings. Cook, covered, in Radarange Oven 2 minutes.

8 ozs. uncooked medium noodles
3/4 cup water

3. Stir in noodles and water. Cook, covered, in Radarange Oven 12 minutes, or until noodles are just tender. Stir halfway through cooking time.

1 cup drained, canned peas

4. Stir in peas. Cool. Wrap tightly and freeze.

1/4 cup grated Parmesan cheese

5. Cook frozen casserole, covered, in Radarange Oven 12 minutes. Then, cook in Radarange Oven using Slo Cook or Automatic Defrost Cycle 10 minutes. Stir mixture. Sprinkle with cheese. Cook, covered, in Radarange Oven 5 minutes, or until hot.

Hamburgers

4 patties to keep in the freezer—either raw or cooked.

1 lb. ground beef
1/2 cup chili sauce
1/4 cup dry bread crumbs
1 teaspoon instant minced onion
1/2 teaspoon salt

1. Combine all ingredients. Divide into four portions. Shape each into patty. Stack, placing double layer of waxed paper between each. Wrap tightly and freeze.
2. Preheat 9-1/2-inch Amana Browning Skillet in Radarange Oven 4-1/2 minutes. Place frozen patties in skillet. Cook, covered, in Radarange Oven 2 minutes. Turn patties over. Cook, covered, in Radarange Oven 7 minutes, or until patties are done.

MICRO-TIP: When patties are cooked before freezing, allow about 6 minutes for heating the frozen patties.

Homemade TV Dinner

Turn leftovers into a TV dinner for 1.

4 ozs. cooked meat
2/3 cup cooked rice, noodles or potatoes
1/2 cup cooked vegetable

1. Arrange foods on heavy plastic or paper plate, placing larger items around edge of plate. Wrap tightly in foil or in freezer type wrap and freeze.
2. Heat frozen dinner, covered with waxed paper, in Radarange Oven 5 minutes, or until steaming hot.

Basic Meat Loaf

Turn a frozen meat loaf into 5 to 6 servings in just over 1/2 hour.

1-1/2 lbs. ground beef
1/2 cup dry bread crumbs
3/4 cup milk
1 egg
1/3 cup chopped onion
1 teaspoon salt
1/2 teaspoon powdered thyme
1/4 teaspoon pepper

1. Combine all ingredients, mixing well. Shape into loaf. Wrap tightly and freeze.
2. Place frozen loaf in utility or loaf dish. Cook, covered with waxed paper, in Radarange Oven 10 minutes. Then, cook in Radarange Ove using Slo Cook or Automatic Defrost Cycle 18 minutes. Rotate dish. Cook in Radarange Oven 5 to 7 minutes, or until meat is set in center

Wild and White Rice Casserole

Rice freezes and reheats nicely. This recipe makes enough for 8.

4 oz. (2/3 cup) wild rice
4 cups water

1 (4 oz.) can mushrooms
1 cup long-grain rice
1 cup chopped celery
1 tablespoon instant chicken bouillon
1 teaspoon salt
1/8 teaspoon pepper
2 tablespoons butter

1. Combine wild rice and water in 2-quart casserole. Cook, covered, in Radarange Oven 8 minutes, or until mixture boils. Let stand at least 4 hours.
2. Drain mushrooms. Stir in with remaining ingredients. Cook, covered, in Radarange Oven 10 minutes, or until mixture boils. Then, cook in Radarange Oven using Slo Cook or Automatic Defrost Cycle 15 minutes, or until liquid is absorbed. Cool. Wrap tightly and freeze.
3. Heat frozen rice in covered casserole in Radarange Oven 25 minutes, or until hot. Stir every 5 minutes.

Baked Rice Deluxe

A company rice dish for 6 that freezes very well.

1 cup chopped onion
2 tablespoons butter

1-1/2 cups quick-cooking rice
1/2 teaspoon salt
1-1/2 cups water

1 (10 oz.) pkg. frozen, chopped spinach
1 (5 oz.) jar sharp process cheese spread
1 (10-3/4 oz.) can cream of mushroom soup
1/4 teaspoon nutmeg

1. Combine onion and butter in 1-1/2-quart casserole. Cook, covered, in Radarange Oven 5 minutes. Stir halfway through cooking time.
2. Add rice, salt and water. Cook, covered, in Radarange Oven 3 minutes Let stand, covered.
3. Cook spinach in Radarange Oven 5 minutes, or until thawed. Drain well. Add to rice along with cheese. Mix well. Stir in soup and nutmeg Cool. Wrap tightly and freeze. Before serving, place frozen mixture in covered casserole. Cook in Radarange Oven 12 minutes. Then cook in Radarange Oven using Slo Cook or Automatic Defrost Cycle 10 minutes. Stir mixture. Cook in Radarange Oven 3 to 4 minutes, or until hot and bubbly.

Autumn Apple Crisp

The orange juice keeps the apples from darkening and also adds a special flavor. Makes 5 to 6 servings.

5 cups sliced apples
1/4 cup orange juice
1/2 cup raisins

1. Combine apples and orange juice in 8-inch round glass baking dish. Mix in raisins.

1/3 cup butter
3/4 cup firmly packed brown sugar
1/2 cup all-purpose flour
1/2 cup quick rolled oats
1/3 cup chopped nuts

2. Heat butter in mixing bowl in Radarange Oven 30 seconds, or until softened. Blend in brown sugar, flour, oats and nuts until crumbly. Sprinkle over apples. Wrap tightly and freeze.

3. Cook frozen dessert, covered with paper towel, in Radarange Oven 16 minutes, or until apples are tender. Rotate dish halfway through cooking time.

Ruby Strawberry Sauce

Rhubarb and strawberries combine for these 5 to 6 servings.

4 cups frozen sliced rhubarb
3/4 cup sugar
2 tablespoons cornstarch
1 (10 oz.) pkg. frozen sweetened, sliced, strawberries

1. Combine all ingredients in 1-1/2-quart casserole.

2. Cook, covered, in Radarange Oven 14 minutes, or until mixture boils. Stir every 3 minutes.

MICRO-TIP: Fresh rhubarb and strawberries may be substituted for frozen strawberries.

Brownies Ala Mode

Brownies, ice cream and chocolate sauce are ready in the freezer for 6 to 8 servings.

1/4 cup butter
1/2 cup sugar
2 eggs
1/2 cup all-purpose flour
1-1/4 cups chocolate-flavor ice cream syrup
1/2 teaspoon vanilla

1. Heat butter in bowl in Radarange Oven 30 seconds, or until softened. Blend in sugar. Beat in eggs. Stir in flour, 3/4-cup ice cream syrup and vanilla. Spread in lightly greased 8-inch round baking dish.

2. Cook in Radarange Oven 5 minutes, or until no longer doughy. Rotate dish halfway through cooking time. Cool. Wrap and freeze. If desired, freeze remaining ice cream syrup in airtight container.

Remaining ice cream syrup
Ice cream

3. Heat frozen brownies in Radarange Oven 2 minutes, or until thawed. Heat remaining syrup in Radarange Oven 1 minute, or until warm. Cut brownies into wedges. Serve topped with ice cream and 2 tablespoons warm syrup per serving.

Notes & Favorite Recipes

As you use your Radarange Oven more you will be developing your own recipes and cooking times. You will want to use this space for a permanent record.

Title and Ingredients	Directions and Times

Notes & Favorite Recipes

As you use your Radarange Oven more you will be developing your own recipes and cooking times. You will want to use this space for a permanent record.

Title and Ingredients	Directions and Times

Notes & Favorite Recipes

As you use your Radarange Oven more you will be developing your own recipes and cooking times. You will want to use this space for a permanent record.

Title and Ingredients	Directions and Times

Notes & Favorite Recipes

As you use your Radarange Oven more you will be developing your own recipes and cooking times. You will want to use this space for a permanent record.

Title and Ingredients	Directions and Times

Index

Index

C

Index

Index

Index

Index

W

Z

Dill Seed
Vegetable juices, yeast breads, deviled eggs, sandwich spreads, lamb roast, creamed chicken, cabbage salad.

Ginger
Cantaloupe, sweet rolls, macaroni and cheese, Indian pudding, carrots, ham sauce, french dressing.

Nutmeg
Tea breads, custard, doughnuts, meat loaf, hard sauce, tomatoes, beans, corn, tomato soup.

Paprika
Beef, veal, chicken salad dressing, barbecue sauce, stuffings, pea soup, cauliflower, corn, garnish.

Herbs and Spices

Flavorings known as spices are truly parts of tropical plants, herbs are always leaves of temperate zone plants. Seeds are actually seeds or fruits of plants grown in either zone.

Storage is important

Store herbs and spices in cool place. Check spice shelf yearly and replace those which have lost aroma.

Parsley
Aspic, biscuits, soufflé omelet, meat pies, seafood salad, remoulade sauce, vegetable soup, carrots, beets.

Sage
Clam juice, corn bread, fondue, veal, fried chicken, baked fish, beets, zucchini.

Savory
Liver paté, scrambled eggs, roast lamb, chicken, green salads, cucumber soup, carrots, beans.

Tarragon
Fruit juice, egg dishes, veal, turkey, potato salad, fish, mushroom soup, beets, zucchini.

Thyme
Aspic, corn bread, roasts, meat loaf, roast duck, tuna, mushrooms, carrots, fish chowder.

Garlic
Used in French and Italian cooking.

Marjoram
Liver paté, sharp cheese spread, scrambled eggs, pot roasts, chicken salad, gravy, onion soup, peas.

Oregano
Chili con carne, pheasant, egg salad, spaghetti sauce, broiled fish, broccoli, tomatoes.

Turmeric
Deviled eggs, curried meats, potato salad, fish sauce, fish, lobster.